NUMERICAL METHODS IN ENGINEERING AND APPLIED SCIENCE:
Numbers are Fun

Mathematics and its Applications

Series Editor: G. M. BELL, Professor of Mathematics, King's College (KQC), University of London

Statistics and Operational Research

Editor: B. W. CONOLLY, Professor of Operational Research, Queen Mary College, University of London

Mathematics and its applications are now awe-inspiring in their scope, variety and depth. Not only is there rapid growth in pure mathematics and its applications to the traditional fields of the physical sciences, engineering and statistics, but new fields of application are emerging in biology, ecology and social organisation. The user of mathematics must assimilate subtle new techniques and also learn to handle the great power of the computer efficiently and economically.

The need of clear, concise and authoritative texts is thus greater than ever and our series will endeavour to supply this need. It aims to be comprehensive and yet flexible. Works surveying recent research will introduce new areas and up-to-date mathematical methods. Undergraduate texts on established topics will stimulate student interest by including applications relevant at the present day. The series will also include selected volumes of lecture notes which will enable certain important topics to be presented earlier than would otherwise be possible.

In all these ways it is hoped to render a valuable service to those who learn, teach, develop and use mathematics.

Mathematics and its Applications

Series Editor: G. M. BELL, Professor of Mathematics, King's College (KQC), University of London

Series continued at back of book

NUMERICAL METHODS IN ENGINEERING AND APPLIED SCIENCE: Numbers are Fun

B. IRONS, B.Sc., D.Sc.
Professor of Civil Engineering
University of Calgary, Alberta, Canada
and
N. G. SHRIVE, M.A., D.Phil.
Professor of Civil Engineering
University of Calgary, Alberta, Canada

ELLIS HORWOOD LIMITED
Publishers · Chichester

Halsted Press: a division of
JOHN WILEY & SONS
New York · Chichester · Brisbane · Toronto

First published in 1987 by
ELLIS HORWOOD LIMITED
Market Cross House, Cooper Street,
Chichester, West Sussex, PO19 1EB, England
The publisher's colophon is reproduced from James Gillison's drawing of the ancient Market Cross, Chichester.

Distributors:

Australia and New Zealand:
JACARANDA WILEY LIMITED
GPO Box 859, Brisbane, Queensland 4001, Australia
Canada:
JOHN WILEY & SONS CANADA LIMITED
22 Worcester Road, Rexdale, Ontario, Canada
Europe and Africa:
JOHN WILEY & SONS LIMITED
Baffins Lane, Chichester, West Sussex, England
North and South America and the rest of the world:
Halsted Press: a division of
JOHN WILEY & SONS
605 Third Avenue, New York, NY 10158, USA

© 1987 B. Irons and N. Shrive/Ellis Horwood Limited

British Library Cataloguing in Publication Data

Irons, Bruce
Numerical methods in engineering and applied science. —
(Ellis Horwood series in mathematics and its applications)
1. Engineering mathematics   2. Numerical calculations
I. Title   II. Shrive, Nigel
511'.02462   TA335

Library of Congress Card No. 86-27660

ISBN 0–7458–0101–3 (Ellis Horwood Limited — Library Edn.)
ISBN 0–7458–0171–4 (Ellis Horwood Limited — Student Edn.)
ISBN 0–470–20803–1 (Halsted Press)

Printed and bound in Great Britain by
Butler & Tanner Ltd, Frome and London

Contents

Bruce Irons, 1924–1983

Foreword

Bruce Irons was one of the inventive giants of 'finite elements'. His name is widely known within the subject area, because of the procedures and innovations he introduced. The Irons 'patch test' is established throughout the literature of engineering and mathematics. Such words as 'isoparametric', 'serendipity', 'frontal solver', and 'semiloof' were introduced to the finite element vocabulary by his fertile imagination, and are now 'household words', providing a memorial to his works.

Bruce Irons was a quiet man with a stroke of genius in his intuitive physical understanding of complex phenomena. He started his engineering career in the Stress Analysis Group of Rolls Royce where many of his innovative ideas germinated—often not understood by his colleagues. Perhaps this lack of understanding was due to his innate modesty which led him to assume that others had equal intellectual abilities. Later, in the atmosphere of university, he learned to transmit his ideas to students by patience and meticulous explanation. Bruce provided outstanding guidance to research workers involved in finite element computer programming, an art in which he achieved a rare degree of perfection. Those who were associated with his teaching and guidance remember him with great kindness and affection. He collaborated widely and taught many who now hold academic posts.

Bruce Irons was fascinated by the seemingly unlimited potential applications of computers to engineering sciences: he was forcibly aware of the dangers associated with misleading computer output resulting from inadequate formulation, erroneous data, or programming errors. The rare blend of theoretical thought, engineering design creativity, computer programming perfectionism, and tutorial practice led him to develop a typical no-nonsense and even ironic 'Irons style'. Those who had the privilege of attending his 1973 lecture in Versailles where he first presented the 'semiloof' concept to the finite element community, will undoubtedly remember his opening slide—displaying a mere egg—followed a few seconds later (that seemed an eternity to the audience) by the provocative statement 'Chickens lay eggs, engineers don't', emphasising the need of any shell element formulation to cope effectively with the corners, edges, and holes met in real-life situations.

His capabilities blossomed, and many of his outstanding works came to fruition during the period spent at University College, Swansea, which he joined in 1966. For his work, Bruce Irons was awarded the DSc of the University of Wales. He was also honoured by the award of the Von Karman prize, jointly with Ian Taig of the British Aircraft Corporation, for the introduction of isoparametric element concepts now used widely in research and industrial finite element codes.

Bruce moved to the University of Calgary in 1974 and devoted his final years to writing books, filling the gaps from earlier research and developing ideas on hybrid elements. He also composed music—another love—and researched computer applications for automatic tuning and harmonising of electronic instruments. Two texts were published during this period. The first *Techniques of finite elements* was written for finite element philosophers, whilst the second, *Finite element primer*, was designed for undergraduate students. Both exhibit many of the famous 'Irons characteristics'.

The manuscript for the present book was begun not long before his untimely death. All those interested in his work and contributions will welcome this publication to complete Bruce Irons's legacy to the engineering community. His last years in Calgary were marred by an incurable illness (multiple sclerosis) which he faced with typical courage. Bruce gathered extensive documentation on the disease and discovered Naudicelle, which gave him remission almost to the end of his life. Indeed, it allowed him to maintain much of his great zest for many years. He is missed by all his colleagues.

[Adapted from the obituary by O. C. Zienkiewicz and Ivan Cormeau that appeared in the *International Journal for Numerical Methods in Engineering*, **20**, 6 (1984), with additions and amendments by Nigel Shrive and Janet Irons.]

Bruce Irons was greatly respected and loved by all in this publishing house, which he often visited. His original proposal was to name the book NUMBERS ARE FUN, which we felt was not a good marketing title for a purely scientific text aimed at engineering and other undergraduate markets; yet the phrase does offer some leavening of mathematical rigour, and, moreover, after discussion with Nigel Shrive, we thought it should be retained because it was Bruce's wish that numbers *can* be fun, and we have therefore incorporated it into the naming of the book as its subtitle.

Ellis Horwood

Preface

Engineering is one of the more numerate professions. In solving problems numerically, both students and practising engineers will want to know which techniques are the most useful, how they work, which can be programmed on calculators, and which on bigger computers. This text is aimed at providing such information; also at indicating potential sources of error in the methods.

This text was in early draft stage at the time of Bruce Irons' death some three years ago. In some respects, it is a memorial to him. I hope it meets the objectives initially set.

Thanks to all those who helped, including: Sue Shrive, Jan Irons, Ivan Cormeau, Ellis Horwood, Carolyn Macarthur (typist), and Fred Sowan (editor of all three of Bruce's books).

Nigel Shrive

1

Fun, frustrations, and tricks of the trade

1.1 AIMS

Practical engineers prefer to solve most problems by number crunching techniques rather than by 'exact' analytical methods. This course is about such 'numerical techniques'. It would be pretentious to call it 'numerical analysis', which is a highly mathematical subject.

Yes, numbers can be fun, mostly. You will laugh at the tricks that numbers sometimes play on us. We will not give any long proofs here, only tiny ones, which may help you to remember something, or which are fun.

Typically, we shall explain a procedure. Then we shall apply it by hand and do experiments, to see how well it performs and what can go wrong. This is not at all frivolous. One day in the real world, this may turn out to be the most useful thing you ever learnt at university.

Numbers are fun. Really. Except sometimes, when there is a deadline or a particularly nasty responsibility, and/or when it's a sum that we can't do properly, and we wish one of our colleagues could take over. A typical practising engineer is calculating, off and on, all the time. Engineering is one of the numerate professions.

Design is calculation. Except sometimes, when you are looking through catalogues for materials or items that are bought outside, finished. Even then, there are usually sums. If you are not doing *any* calculations, then, face it, you are copying; or you are functioning as an artist, in the wrong trade union.

Rough sums. A good habit—15 digit accuracy doesn't matter; except sometimes, as we shall see, when your final answers are abnormally sensitive to small errors in your intermediate numbers. Very often, you can do a rough sum, with terrible assumptions, then another with different assumptions, and so on, then guess nearly the right answer. A good habit. Rough sums are often done in a frenzied, haphazard way in one's head, on the back of an envelope. Such speculation will allow you to discard many options. Indeed you will probably come to agree, when you have to take responsibility, that it isn't usually the rejected options that get you into trouble. It's the sums that you simply *forget* to do—aspects that never occurred to you. An engineer should think of every possibility!

Again, finance. An engineer sometimes 'blows the whistle' to the accountants, who understandably don't want to become involved in speculative, approximate costing. For a start, how much does typing a letter cost? ($20, £10?) Hand calculations? (5 per minute on a hand calculator is fairly good: about 6 p(c) each.) A training programme? A meeting? (Guess the salaries of all the other people present. Guess the overheads. Actually do the sum. Was a particular meeting justified, costwise?) A Fortran compiler with poor diagnostics? A computer program? (About $20, £10 per line of Fortran. A full-time programmer expects to write, and verify, and document about 20 lines per day—and maintain existing programs, and answer queries, etc.) A badly written computer program? Badly written manuals? An unfriendly computer? Accountants should try to cost all such things, but don't dare. Nobody really knows, in many cases. Do rough sums—you may hit on something vitally important to your firm.

1.2 TOOLS

A modern numerical methods course is about *electronic* computing. This makes all the difference. Before, the typical engineer relied on his slide rule, which was very tiring to operate for more than an hour or two. An engineer who was expected to calculate most of the day, much as a secretary is expected to type, was usually equipped with a machine about as large as a typewriter, sometimes with a motor, usually with a handle that had to be turned. It could add, subtract, multiply, and divide, and take square roots (with some difficulty). These machines were affectionately called 'rabbits', because they multiplied remarkably quickly—seconds, as against minutes by hand! But they were less tiring to use than slide rules. Today your kid sister would be infinitely better equipped!...

1. *Pocket calculators.* The greatest innovation of the decade 1970–80 was the programmable pocket calculator. You should never be without one. If yours is programmable, but you haven't yet learnt to write programs for it, then learn *now*. Much better than computers for simple, everyday jobs. No hassles. Engineers use calculators all the time. Small is beautiful ... and less trouble!

2. *Micros, desktop computers.* Better than terminals: easier to use, no interruptions, no logging in, no cost-per-minute, immediate response, sufficient memory and speed for most jobs, 'portables' with flat screens to fit into your briefcase. But they will never fit into your pocket. Usable screens and keyboards obviously can't be miniaturised. They are potentially very power-

ful; through 'networks' they can talk to the most powerful computers available....

3. *Supercomputers.* Performance measured in 'Megaflops' (per second). A single 'floating point operation' (flop) is, e.g. $2.718\,281\,843\,263\,511/3.141\,592\,653\,589\,793 = \ldots$. The current norm (1985) is 20–800 megaflops. Towards 1990, we can expect to see special purpose supercomputers capable of 100 gigaflops every second (10^{11} flops), for example to analyse pictures from satellites, i.e. about a thousand times faster still. (At present, they don't use special chips, just cleverly designed circuits working in parallel.) But rather expensive...

4. *Array processors?*—The poor man's supercomputer? Since the very early days of computing, Grosch's law has been more or less honoured: (power) α(expenditure)^2. In this cut-throat market, however, there are spectacular perturbations, commercial accidents. (Things go mad for a while.) Array processors for example. They can't work alone; they need an ordinary computer to control them. Unlike supercomputers, their memories are not large enough at present for certain 'superjobs'. But they cost about as much as 'minicomputers'....

5. *Minicomputers?* An earlier perturbation from Grosch's law, starting about 1970 with the Data General 'Nova'. 'Mini' is hardly the right word today. You can get a machine that accepts programs with 400 000 lines of Fortran! But don't be misled, as so many people are, by the idea of your own departmental mini. It sounds marvellous until you start budgeting for an operator and/or a systems programmer, to keep it running! People are expensive. Better to have the Computer Centre pay their salaries where possible....

6. *A Centralised Computer Complex?* It is usually a good tradeoff to give up some of your independence, to save money. Your own personal computer should provide enough independence in practice, provided you can overflow at any time onto a central computer, for extra number-crunching, and backup memory for extra security. (A good combination is to have pre-programs, which generate data on your own micro, to send to the big computer. Buzzword, 'work station'.) Your firm will probably have a Computer Centre. Or perhaps supercomputer complexes many miles away will become the norm—or super-supercomputers? With networks of optical fibres? Prospects are mind-boggling.

1.3 NUMBER-CRUNCHING, EXPERT COMPUTERS

The effects of the computer revolution have hardly begun, but already these are exciting days. Everything will be subtly different, with supercomputers everywhere. In 1950 we regarded certain sums as large—they are now regarded as trivial, commonplace. In the same way, today's mammoth calculations will soon be regarded as kid's stuff. The supercomputers, when they get cheap enough, will create their own demand! In design, there will be far fewer experiments with models—wind tunnel tests for example. Already, if a company must either cut costs, or slowly die, there is a frantic drive to put everything possible on the computer—already this is usually cheaper than the corresponding experiment. Many more things will go into production almost off the drawing board. Already computers are largely designed by computers!

Decisions are more difficult than numbers, of course. 'Artificial Intelligence' is in its infancy. However, nothing is intrinsically impossible for computers, which can already beat most grand-masters in chess—admittedly not so much by outwitting them, as by wearing them down, and making fewer mistakes!

1.4 LITTLE SUMS

In your own work; try never to *rely* on computers. Check the answers. Engineering will always depend on simple little sums—like those you do in your courses. Back-of-the-envelope sums, crude ballpark estimates. At most, pocket calculator stuff. You just have to get these right! The marking scheme in a typical quiz might be:

Method 5/10

Correct answer 3/10

Neatness 2/10

This may come as a shock to you: In the real world you will be expected to complete the arithmetic—and to get the right answer. Otherwise some person senior to yourself will be using your guts for garters. *Always* write down your dimensions as a check, and cancel them as in fractions. A trivial example:

Distance travelled in kilometres		Unit cost	
	miles/km	$/litre	
200	× 0.625	× 0.50	
Cost = —————————————————————			
40	× 0.23	× 1.85	
miles/gallon		gallons/litre	$/£

= £3.67

For example, delete 'miles' in miles/km above the long line, and in miles/gallon, below the long line, just as you would if you had identical numbers in a complicated fraction. Do the same with $ and litres. Cancel 'gallon' in miles/gallon and in gallons/litre. Cross out 'Kilometres'. You are left with '/£' under the long line, i.e. £. This completes your dimensional check. Start the habit, today. It is not only in electricity, mechanics, and strength of materials that you can apply this check, to avoid the painful consequences. Your face will never be redder, than when you are found to have multiplied, instead of divided. Everybody will say, 'Didn't you notice that? Didn't it *look* wrong? Tell that to the Bishop!' The trouble is, you are entering a peculiarly *numerate* profession—you will be expected to know, almost by instinct, the order of magnitude of all the quantities that you have regular dealings with. But it isn't really an instinct at all. It's the cumulative experience of having done countless rough sums over the years.

1.5 NUMERICAL METHODS

The bulk of this course is not about order-of-magnitude check calculations, important as these are. We leave that to other courses—although we know in fact that it is a lesson you will have to learn 'on the job'. It is about the kind of sum that you will do in industry, trivial for the computer, but by hand, with pencil and paper, it would take forever.

This course is predominantly about the interaction of computers and the profession of engineering, at the most basic technical level. Times are changing. Yet engineering students are often heard to complain, 'Why are we having to learn Fortran? Aren't there armies of overpaid(?) programmers everywhere in industry? You just tell them what you want, in words of one syllable, and they go away and do it. Like secretaries.'

Oh, dear. The short answer is that Fortran was invented so that engineers could do their own programming, and take full responsibility, most of the time. Then the program will—hopefully!—do all that the engineers want. That is, if the job is not too difficult, or too demanding; if the program is expected to fit into a company-wide system, for example, then it may be asking too much of a typical engineer.

The longer (and more controversial) answer is that wise management will increasingly hand out programming jobs to the people who are more interested in the answers than in the programming techniques. Career programmers are smart people, but they tend to be loyal to their discipline, like doctors are to their profession, rather than to the company that happens to be paying their salaries. They sometimes produce unnecessarily sophisti-cated programs, etc. They would regard themselves as a part of the programming team with technical programming standards to maintain, rather than the engineering team, or the accounting team, etc. We can sympathise, because programming is difficult.

It is so difficult in fact that if you, as a manager, were to give a tough programming job to a sufficiently smart engineer, then he would cease to be an effective engineer, for a year or so, and you would lose the benefit of his smart engineering ideas. So Fortran isn't always such a good thing. Posh or difficult programming is best done by professionals.

Like typing. Modern word processors can produce book-style docu-ments, if and when such high quality is justified. Come the revolution, day-to-day typing will perhaps be done by somebody junior, working at breakneck speed, and you, the engineer, will be expected to make all the corrections and alterations yourself, on a word processor! Or perhaps you will have to intone each word carefully into a special microphone and thus avoid using a typist altogether! The advertisements won't tell you that sort of thing. Life will be—different. Secretaries will be scarce, expensive, and very, very, beautiful!

Times are changing, as never before. We are only guessing at the details. But about the importance of this course throughout your future career in a computerised world, we are not guessing. Before we start, we must convince you that computer-sized sums really are worth doing. Otherwise you won't be motivated to treat this course seriously. Computer Modelling.

The stresses in the crankshaft of your car, for example. Or the pollution that you predict in the water supply for a town, when a particular factory is built. Or, to take a simple example: a mountain in central Australia is composed of high-grade iron ore. Determine how much. Now there is no formula for the volume of a mountain! You must survey it carefully, then estimate the volume, from the numbers. You have a numerically defined surface, a 'numerically defined function'. That is a good start. Bear it in mind.

EXAMPLES

1. Estimate roughly the size of a 100 000 tonne ship.

2. Estimate roughly how much your class is worth per hour, from the earning power of the students. Guess how much your teacher earns. Compare these, and discuss. Guess the overheads, including administration.

3. Discover the cost/rental/maintenance of your computer. Guess the staff salary and overheads. Compare these, by putting them on a 'per annum' basis. Would you prefer to spend more on the hardware/on the staff? ('Both' is not an acceptable answer.)

2

The Taylor series, etc.

2.1 OVERVIEW

Let's start the course. This chapter is more academic than most of what follows; indeed, more than we would wish. However, the Taylor series is an important idea that can't really be popularised. It extends a basic notion that a person with no technical training will find obvious. Suppose we are driving up a 10:1 gradient. Each time we proceed 100 metres horizontally we expect to ascend 10 metres (see Fig. 2.1). This assumes however that the slope remains 10:1. If the slope varies, then it is less obvious what will happen. It is this more general problem that Taylor addresses.

Let us express these ideas algebraically. Let x be the horizontal distance travelled, and let $f(x)$ be the height above sea level. Then $df(x)/dx = 0.1$ and the increased height on travelling a further horizontal distance h is $0.1\,h$, or in general,

$$f(x_0 + h) - f(x_0) = h\frac{df}{dx} \dagger$$

or

$$f(x_0 + h) = f(x_0) + h\frac{df}{dx}. \qquad (2.1)$$

Let us examine this further, in order to tie together the intuitive and the algebraic concepts. Differentiating (2.1) with respect to h gives $df/dh = df/dx$. This makes complete sense in the case with $df/dx = $ constant;

---

† If you divide both sides by h, then consider the left-hand side; you should recognise the standard definition of a derivative at the limit of h tending to zero.

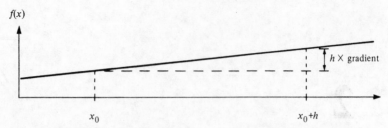

Fig. 2.1 A car ascending a hill of moderate, uniform gradient: an introductory example for the Taylor series.

$df(x_0 + h)/dh$ is equal to the same constant, identically. Thus, $f(x_0 + h)$ and the expression (2.1) must be identical. In Fig. 2.1 we had the slopes the same at every point (0.1) and putting $h = 0$ we make the values the same as x_0.

Taylor does this, in a more sophisticated way. Our purpose in starting the course with Taylor's series is

1. To lay sturdy foundations; the series is constantly in use.
2. To give intuitive plus graphical insights. Mental pictures remain, long after the detailed mathematical arguments have slipped from memory. We are engineers!
3. To plant the concept of using polynomials to represent smooth functions.
4. To introduce the idea that it is usually surprisingly accurate to take the mean slope as the slope halfway along the region considered.
5. To touch upon Taylor's theorem, which is the mathematical basis of nearly all of the 19th century mathematics that we use in engineering.
6. At the end of the chapter, to gain some real practical experience when things start going wrong!

Now, Taylor's series extends (2.1) indefinitely:

$$f(x_0 + h) = f(x_0) + hf'(x_0) + \frac{h^2}{2!}f''(x_0) + \cdots \qquad (2.2)$$

where we use f' as an abbreviation for df/dx and f'' for d^2f/dx^2—you've probably seen that before.

We usually use Taylor for estimating the value of $f(x_0 + h)$ when h is small. Figure 2.2 explains.

However, in some instances Taylor is amazing: h can be any size. For a simple example, consider $f(x) = \sin(x)$, x in radians, and put $h = x$, $x_0 = 0$ in equation (2.2). You have probably seen the series for $\sin(x)$ already: Equation (2.2) now matches it.

$$\sin x = x - \frac{x^3}{3!} + \frac{x^5}{5!} - \frac{x^7}{7!} + \cdots. \qquad (2.3)$$

This series converges, by the ratio test: however large x is, the ratio, $(N + 1)$th term/Nth term, tends towards zero with large N. (If this ratio r lies between -1 and $+1$, the strict mathematical argument proceeds by comparing the

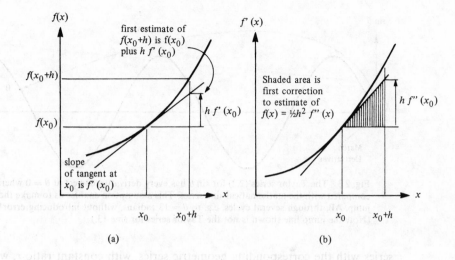

(a)

(b)

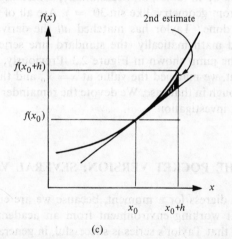

(c)

Fig. 2.2 The usual application of Taylor. We know about the curve $f(x)$ at $x = x_0$. We wish to estimate the value of $f(x)$ at $x = x_0 + h$. Our first estimate in (a) involves $f(x_0)$ and the slope at $x = x_0$: Taylor's first two terms as we saw before (Figure 2.1). We end up a bit short because the slope $f'(x)$ changes with x, so we plot that in (b). We then estimate in (b) how much the slope has changed. The correction to the change in $f(x)$ is the shaded area (the change in f is the integral of $f'(x)\,dx$, if you like). This is Taylor's third term, and is transferred to the x vs $f(x)$ diagram in (c)—a quadratic addition to our previous straight line. Well, we realise from (b) that $f''(x)$ changes with x, so now we should make another correction. If we drew it, the triangular area on the x vs $f''(x)$ diagram would show as a quadratic addition to (b), with height $(\frac{1}{2}h^2\ f'''(x_0))$ at $x_0 + h$. The area under the parabola will be $(1/3)h(\frac{1}{2}h^2\ f'''(x_0))$ ((1/3) base x height) since the vertex will be at $x = x_0$. The term is recognisable as Taylor's 4th term, $(1/3!)h^3\ f'''(x)$. And so on, ad infinitum until you're happy with your estimate (or bored!). But *note*: We can substitute $h = x - x_0$ in our Taylor series, which is useful if we expand the series about a non-zero value of x_0.

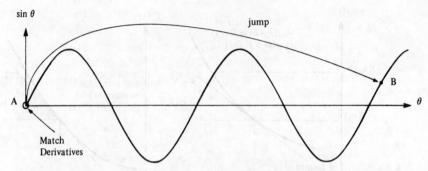

Fig. 2.3   The Taylor series (2.2) for sin θ has every derivative correct at $\theta = 0$ when compared with the mathematical sine series. This assumption enables us to make the jump, AB, through several cycles, e.g. to $\theta = 13$ radians, without introducing error! (Note the jump line shown is not the Taylor series for sine 13.)

series with the corresponding geometric series, with constant ratio r, which always converges if $-1 < r < 1$.)

Thus we have (2.3), which since it converges represents a definite function, and we have 'sin x' proper, whatever that means: an abstract mathematical function, or the result of pressing a button on your calculator, or something you look up in tables, or, occasionally, something that your kid sister could argue from geometry, like sin $30° = \frac{1}{2}$. Are all of these the same? What has Taylor done? Taylor has matched *all* the derivatives from (2.2) to those deduced mathematically (the standard sine series) at $x = 0$. Thus we can make the jump shown in Figure 2.3. Previously, for the case with $df/dx =$ constant, we matched the value at $x = x_0$ and the slope everywhere, which was enough in that case. We devote the remainder of the chapter to this more difficult investigation.

2.2   THE POCKET VERSION: SEVERAL VARIABLES

But we digress for a moment, because we are engineers, with a somewhat different working environment from an academic mathematician. Let us assume that Taylor's series is successful, in general. The engineer's recurrent problem is in remembering it correctly, and in applying it to complicated cases. A useful formula is

$$f(x_0 + h) = e^{h\,d/dx}f(x_0)$$

$$= \left(1 + h\frac{d}{dx} + \frac{h^2}{2!}\frac{d^2}{dx^2} \cdots \right)f(x_0) \qquad (2.4)$$

which is (2.2) again. All you have to remember is the exponential series. We shall apply this artificial 'operator' to great effect in deriving strange and useful formulae.

But our immediate purpose in presenting (2.4) is to extend the Taylor series to several variables, in an easily remembered form. First, we must remind you what 'partial derivatives' mean. If x points East, and y points North, and z points up, then we must use ∂ instead of d. Thus $\partial z/\partial x$ is the slope of the hill, if we climb it in an easterly direction, keeping y constant. Again, $\partial z/\partial y$ is the northward slope, with x constant. (Please don't feel insulted, if you knew this already.) Using partial derivatives, Taylor's series in two variables can be written

$$f(x_0 + h, y_0 + k) = e^{h\partial/\partial x + k\partial/\partial y}f(x_0, y_0)$$

$$= f(x_0, y_0) + \left(h\frac{\partial}{\partial x} + k\frac{\partial}{\partial y}\right)f(x_0, y_0)$$

$$+ \frac{1}{2!}\left(h\frac{\partial}{\partial x} + k\frac{\partial}{\partial y}\right)^2 f(x_0, y_0) + \text{etc.} \qquad (2.5)$$

For example, the cubic terms are given by

$$\frac{1}{3!}\left(h\frac{\partial}{\partial x} + k\frac{\partial}{\partial y}\right)^3 f(x_0, y_0) = \frac{h^3}{6}\frac{\partial^3 f(x_0, y_0)}{\partial x^3} + \frac{3h^2 k}{6}\frac{\partial^3 f(x_0, y_0)}{\partial x^2 \partial y}$$

$$+ \frac{3hk^2}{6}\frac{\partial^3 f(x_0, y_0)}{\partial x \partial y^2} + \frac{k^3}{6}\frac{\partial^3 f(x_0, y_0)}{\partial y^3}.$$

Thus if $f(x, y) = xy^2$, and $x_0 = y_0 = 0$, you can easily check that there is only one non-zero cubic term in Taylor's series, namely hk^2. But if we had not given you the pocket formula, we assure you that it would have wasted a lot of your time re-inventing it. This handy formula might have been designed specifically for busy engineers!

2.3   APPROXIMATE TERMINATION

We postpone Taylor's *theorem*, which sometimes gives mathematical error bounds when the series is terminated. Frankly, this is seldom of urgent interest to busy engineers; but the first stage by which we introduce the proof here leads to a simple, almost obvious, satisfying, and immensely useful rule of thumb: mean slope $\cong$ midway slope. See Fig. 2.4. Let us start with a simple example, $f(x) = x^2$. Thus $f(7) = 49$, $f(9) = 81$, so that the mean slope is $(81 - 49)/2 = 16$. The exact slope $f'(x) = 2x$, so the slope midway between 7 and 9, at $x = 8$, is 16. Indeed, the rule of thumb is *exact* for any quadratic: mean slope = midway slope.

Let's investigate a more general case, and let's do it properly. Let's terminate the Taylor series at any term, say the fourth. For convenience, we no longer use x_0:

$$f(x + h) \cong f(x) + hf'(x) + \frac{h^2}{2!}f''(x) + \frac{h^3}{3!}f'''(x)$$

$$\cong f(x) + hf'(x) + \frac{h^2}{2!}\left\{f''(x) + \frac{h}{3}f'''(x)\right\}.$$

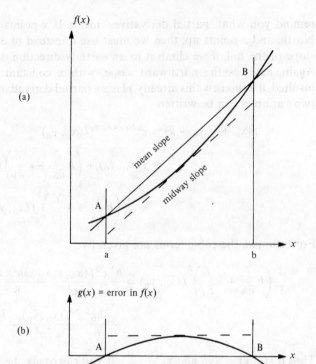

Fig. 2.4   (a) The chord gives the mean slope along interval AB. By eye, this is nearly parallel to the tangent halfway along the interval. (b) In particular, $g(x)$ is zero at A and B. What goes up must come down: and between going up and coming down it must be stationary—if the slope exists and if the slope is continuous everywhere. If the mean slope along AB is zero, the actual slope must be zero somewhere between A and B. Obvious. (Although mathematicians don't like to hear you say that.)

Now let us make a bold assumption. Let us assume that the first three terms (to h^2) give 99 percent of the correct answer; also that the following term supplies, almost exactly, the remaining 1 percent needed. This, or something like it, is a very common situation. We now interpret the expression in curly brackets as the start of a subsidiary Taylor series:

$$f''(x + \tfrac{1}{3}h) \cong f''(x) + \frac{h}{3}\frac{d}{dx}f''(x)$$

$$\cong \left\{ f''(x) + \frac{h}{3}f'''(x) \right\},$$

i.e. by recognising the item in curly brackets as a miniature Taylor series in its own right (which introduces a second approximation) we give the original series a new and intriguing interpretation:

$$f(x + h) \cong f(x) + hf'(x) + \frac{h^2}{2!}f''\left(x + \frac{h}{3}\right). \qquad (2.6)$$

Obviously, we did not have to terminate at the third term in (2.6). If we had terminated at h^n, i.e. at the nth derivative, and if we had made similar assumptions, we should have to find the nth derivative not at x, as used in the previous terms, but as $x + h/(n + 1)$. Try it with another n.

This approximate theory is surprisingly accurate, usually. Try it and see. For example, with $n = 4$,

$$e^{0.5} \cong 1 + 0.5 + \frac{(0.5)^2}{2!} + \frac{(0.5)^3}{3!} + \frac{(0.5)^4}{4!} e^{0.1}$$

using the approximate theory (eq. (2.6)) with $x = 0$ and $h = 0.5$. To get the exact answer would have required $e^{0.10343}$ in the last term. This example was chosen for simplicity.

For practical purposes we shall concentrate on the simplest case, with $n = 1$:

$$f(x + h) \cong f(x) + hf'\left(x + \frac{h}{2}\right)$$

or

$$f'\left(x + \frac{h}{2}\right) \cong \frac{f(x + h) - f(x)}{h} \tag{2.7}$$

i.e. midway slope $\cong$ mean slope.

Honestly, this rule of thumb was worth the laborious development that we have given it: look at Fig. 2.4 again. For it takes us into the next topic, which is very useful in practice.

2.4   FINITE DIFFERENCES

We now formalise the crude rule, midway slope $\cong$ mean slope, into a sophisticated tabular layout. Finite difference theory leads to fascinating mathematical abstractions, if that is the direction you wish to take. There is even a 'calculus' of finite differences. We use it later in this book. You will be expected to know something about it. But for engineers who use finite difference techniques with the slightest understanding, the main key to their 'understanding' of the method is usually our approximate mean slope rule.

Let's see an example of the 'finite difference table':

x	$f(x) = e^x$	δf	$\delta^2 f$	$\delta^3 f$
0.7	2.014			
		0.212		
0.8	2.226		0.022	
		0.234		0.002
0.9	2.460		0.024	
		0.258		
1.0	2.718			

Explanations: $0.212 = 2.226 - 2.014$, $0.234 = 2.460 - 2.226$, $0.258 = 2.718 - 2.460$, $0.022 = 0.234 - 0.212$, etc. Note, successive columns use interlacing levels; when you write a number which is the difference of two numbers to the left, you place it at a level midway between them. Midway slope... This wastes a little paper, but it is unwise to disregard the rule.

The level corresponds to a value of x. Thus we say that 0.212 is $\delta f(0.75)$ where δ is the 'central difference operator' $(\delta f_{i+1/2} = f_{i+1} - f_i)$, the difference of the two values in the previous column: $0.022 = \delta^2(0.8)$—we are back to the level of $x = 0.8$. Often the f is dropped, and thus δ has two obvious meanings:

1. Move one space to the right in the table.
2. In this column to the right, put the difference of two sequential values in the previous column (at a level half way between them). By the mean slope rule, δ should be roughly equivalent to $h\, d/dx$ calculated at the same level of x. h is the difference in x values ($h = x_{i+1} - x_i$). Let's test the idea:

$$\text{For } x = 0.75, \qquad \delta f(0.75) = 0.212.$$

Because $h = 0.1$ and $f(x) = e^x$ in the table,

$$h\frac{df}{dx} = he^{0.75} = 0.21170 \qquad \text{This is in good agreement.}$$

For $x = 0.8$, $\delta^2 f(0.8) = 0.022$

$$\left(h\frac{d}{dx}\right)^2 e^x = h^2 e^{0.8} = 0.02226.$$

This is more useful than would appear. From

$$\delta = h\frac{d}{dx}$$

it should follow that

$$\frac{d}{dx} = \frac{1}{h}\delta \tag{2.8}$$

which compares with equation (2.7)

$$\frac{d}{dx}f\left(x + \frac{h}{2}\right) = \frac{1}{h}\left(f(x + h) - f(x)\right)\left(= \frac{1}{h}\delta f\right).$$

Thus we should have a means for finding derivatives. We demonstrate this by defining δ in a third more mathematical way:

3. If E is another operator which means 'one more space down in the table', then because $\delta f(0.75) = f(0.8) - f(0.7)$, it follows that there is a general relation between the two operators:

$$\delta = E^{1/2} - E^{-1/2} \tag{2.9}$$

$E^{1/2}$ means 'take the value in a column as $\frac{1}{2}$ space below the level of x under consideration'.

$$\delta^2 = (E^{1/2} - E^{-1/2})^2 = E^{+1} - 2 + E^{-1}$$
$$= f(x + h) - 2f(x) + f(x - h)$$

Note E^{+1} actually is E, given our definition. The 'calculus of finite differences' really starts with this relationship. We can now treat the same example in a more sophisticated way. From (2.8) we can say

$$\frac{d^2}{dx^2} = \left(\frac{d}{dx}\right)^2 = \left(\frac{1}{h}\delta\right)^2.$$

Thus to find the second derivative of our function $f(x) = e^x$ at $x = 0.8$.

$$\frac{1}{h^2}\delta^2 f(0.8) = \frac{1}{0.01}(E - 2 + E^{-1})f(0.8)$$

$$= \frac{1}{0.01}\{f(0.9) - 2f(0.8) + f(0.7)\}$$

$$= 2.227$$

carrying as many places as are retained naturally on our calculator. In fact, $d^2f(0.8)/dx^2 = \exp(0.8) = 2.226$, so we have a simple and often very accurate formula. The second difference as shown here is an especially favourite device. Let's look at some things we can do with finite differences.

First Example
For $f(x) = \tan x$ use central difference operators to find $f'(1)$ and $f''(1)$ with $h = 0.2$. Check the estimates with smaller values of h and the actual value obtained by differentiation.

Solution
To find f' using central difference operators, we use the 'mean slope rule':

$$f'(x) = \frac{f(x + h) - f(x - h)}{2h}$$

and the second derivative is

$$f''(x) = \frac{f(x + h) - 2f(x) + f(x - h)}{h^2},$$

so with $h = 0.2$

$$f'(1) = \frac{\tan(1.2) - \tan(0.8)}{0.4} = 3.85628$$

with

$h = 0.1$ $f'(1) = 3.52301$ —difference of $0.33—10\%$ from previous value

$h = 0.05$ $f'(1) = 3.44933$ —difference of $0.07—2\%$ from previous value

$h = 0.01$ $f'(1) = 3.42646$ —difference of $0.02— <1\%$ from previous value

Actual Value

$$f'(x) = \frac{1}{\cos^2 x} \qquad \therefore \; f'(1) = 3.42552.$$

For the second derivative with $h = 0.2$

$$f''(1) = \frac{\tan(1.2) - 2\tan(1) + \tan(0.8)}{(0.2)^2} = 12.17437$$

with $h = 0.1$ $f''(1) = 11.01024$

$\quad\ h = 0.05$ $f''(1) = 10.75298$

$\quad\ h = 0.01$ $f''(1) = 10.67316$

Actual Value

$$f''(x) = \frac{2\sin x}{\cos^3 x} \qquad \therefore \; f''(1) = 10.66986.$$

COMMENT

We can see that the choice of $h = 0.2$ is too inaccurate, that the smaller intervals give better results. Normally where we don't have an explicit solution to check against, it is sensible to check with smaller intervals: otherwise you have no check on the accuracy of the estimate.

Second Example

Given the following values of x and $f(x)$, what does a finite difference table tell you about $f(x)$?

Given		we calculated		
x	$f(x)$	δf	$\delta^2 f$	$\delta^3 f$
1	4.000			
		−0.073		
1.1	3.927		−0.158	
		−0.231		−0.018
1.2	3.696		−0.176	
		−0.407		−0.018
1.3	3.289		−0.194	
		−0.601		−0.018
1.4	2.688		−0.212	
		−0.813		−0.018
1.5	1.875		−0.230	
		−1.043		
1.6	0.832			

Conclusions: We note $f'''(x)$ is a constant, hence the highest 'power' in $f(x)$ is a cubic.

Now, since $\delta^3 = h^3(\mathrm{d}/\mathrm{d}x)^3$ and $h = 0.1$ then, $(\mathrm{d}/\mathrm{d}x)^3 = -18$ from our table.

Now, $(\mathrm{d}^3/\mathrm{d}x^3)\cdot x^3 = 6$, so we can conclude that $f(x)$ contains a $-3x^3$ term as the highest order.

We could now remove $h^2(\mathrm{d}^2/\mathrm{d}x^2)$ operating on $(-3x^3)$ from the $\delta^2 f$ column; i.e., remove $-0.18x$). Thus $\delta^2 f$ (1.1) would become $-0.158 + 0.18$ (1.1) $= 0.04$. We would find this constant with x which would tell us that $f(x)$ also contains a square term in x. We could also proceed to determine what it was. Maybe you would like the exercise.

2.5 TAYLOR'S THEOREM

The theorem is pukka mathematics, based on the same general idea as the approximate termination discussed in Section 2.3:

$$f(x + h) \cong f(x) + hf'(x) + \frac{h^2}{2!}f''(x) + \cdots + \frac{h^n}{n!}f^{(n)}\left(x + \frac{h}{n+1}\right)$$

but introducing exact equality. With approximate termination, we computed the final derivative at a point a fraction $1/(n + 1)$ along the jump; according to Taylor's theorem, the terminated series is exact, if we evaluate the final derivatives at some *unknown* point *within the jump*:

$$f(x + h) = f(x) + hf'(x) + \frac{h^2}{2!}f''(x)$$

$$+ \cdots + \frac{h^n}{n!}f^{(n)}(x + \xi h) \qquad \text{where } 0 < \xi < 1 \ldots \quad (2.10)$$

We shall prove this for $n = 4$, by writing the coefficient of $h^n/n!$ as the unknown A in an equation

$$f(x + h) - f(x) - hf'(x) - \frac{h^2}{2!}f''(x) - \frac{h^3}{3!}f'''(x) - \frac{h^4}{4!}A = 0.$$

For argument, we replace h by v, thus defining a new function, as illustrated in Fig. 2.5:

$$r(v) = f(x + v) - f(x) - vf'(x) - \frac{v^2}{2!}f''(x) - \frac{v^3}{3!}f'''(x) - \frac{v^4}{4!}A.$$

Our first equation for A now becomes simply $r(h) = 0$. Also $r(0) = 0$, and by differentiating the formula above; $r'(0) = r''(0) = r'''(0) = 0$.

Differentiating $r(v)$ a fourth time,

$$r''''(v) = f''''(x + v) - A = w(v) = 0, \qquad \text{when } v = v_4.$$

By the graphical argument in the caption to Fig. 2.5, $0 < v_4 < v_3 < v_2 < v_1 < h$. Thus we have proved Taylor's theorem, by devious commonsense, and by drawing pictures; $A = f''''(x + v_4)$. This is considered a rather difficult theorem, and it really is difficult to grasp in mathematical jargon. Figure 2.5 should help.

Our example confirms Taylor's theorem numerically.

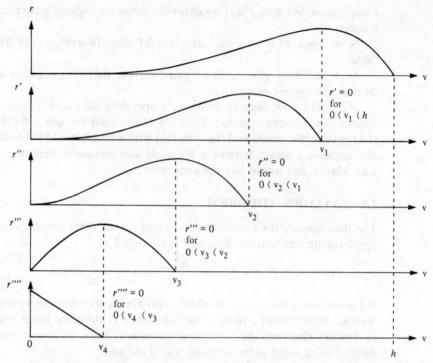

Fig. 2.5 At $v = 0$, r and its derivatives r', r'', and r''' are zero. At $v = h$, $r = 0$, as required. We apply the argument of Fig. 2.4(b) four times in succession: r, r', r'', or r''' is zero at both 'ends'. In each case, what goes up must come down; so there must be a horizontal region somewhere between, where the derivative is zero. The conclusions are summarised thus: $h > v_1 > v_2 > v_3 > v_4 > 0$, which means that $0 < \xi < 1$, where $\xi^1 = v_4/h$.

We estimate $f(x) = 1/(1 + x)$ at $x = 0.5$, when the series is expanded about $x = 0$ and is terminated at the fourth term (i.e.: $n = 3$):—

$$f(x_0 + h) = f(x_0) + hf'(x_0) + \frac{h^2}{2!}f''(x_0) + \frac{h^3}{3!}f'''(x_0 + \xi h)$$

where $0 < \xi < 1$, remember?

Now

$$f(x) = \frac{1}{1 + x} \qquad \text{with } x_0 = 0, h = x; \qquad \text{so } f(x_0) = 1$$

$$f'(x) = \frac{-1}{(1 + x)^2} \qquad \begin{array}{l}\textit{(N.B. If we expand about a}\\ \textit{different point, say } x = 1,\\ \textit{then } h = x - 1).\end{array} \qquad hf'(x_0) = -x$$

$$f''(x) = \frac{2}{(1 + x)^3} \qquad\qquad\qquad\qquad\qquad h^2f''(x_0) = 2x^2$$

$$f'''(x) = \frac{-6}{(1 + x)^4} \qquad\qquad\qquad\qquad\qquad h^3f'''(x_0 + \xi h) = \frac{-6x^3}{(1 + \xi x)^4}$$

hence, the terminated Taylor expansion is

$$f(x) = 1 - x + x^2 - \frac{x^3}{(1 + \xi x)^4}$$

with

$$x = \frac{1}{2}, \quad f(x) = \frac{1}{1 + \frac{1}{2}} = \frac{2}{3}$$

and by the series

$$f(x) = 1 - \frac{1}{2} + \frac{1}{4} - \frac{1}{8(1 + \frac{1}{2}\xi)^4}$$

$\therefore$ to determine ξ

$$-\frac{2}{3} + 1 - \frac{1}{2} + \frac{1}{4} = \frac{1}{12} = \frac{1}{8(1 + \frac{1}{2}\xi)^4}$$

$$\therefore \quad 1 + \frac{1}{2}\xi = \sqrt[4]{1.5}$$

$$\therefore \quad \xi = 0.213$$

which indeed lies between 0 and 1.

Does Taylor's theorem have any practical use, since we don't know the value of ξ beforehand? Sometimes, for example:

What is the largest angle which will give a relative error of at most 1×10^{-2} in the approximation $\sin \theta = \theta$; what is the actual relative error for this angle?

One solution is to apply Taylor's theorem to $\sin \theta$, terminating at the third term,

$$f(x + h) = f(x) + hf'(x) + \frac{h^2}{2!}f''(x + \xi h)$$

Set $x = \theta_0 = 0$ so $h = \theta$, and $f = \sin$.

$$\sin \theta = \sin (0) + \theta \cos (0) - \frac{\theta^2}{2!} \sin (\xi\theta)$$

$$\sin \theta = \theta - \frac{\theta^2}{2!} \sin (\xi\theta)$$

Now, since $0 < \xi < 1$, for small θ,

$$\left| \frac{\theta^2}{2!} \sin (\xi\theta) \right| \leq \frac{\theta^2}{2!} \sin \theta$$

that is, $\theta^2/2! \sin \theta$ is the biggest error we could expect. We want the relative error with respect to θ to be less than 10^{-2}.

i.e. $\quad \dfrac{\frac{\theta^2}{2!} \sin \theta}{\theta} < 10^{-2}$

which gives $\theta = 8° 6' 58'' (= 8.1161°)$ for the actual error,

$$\sin \theta = 0.14117961$$

$$\theta = 0.14165268$$

so the error = $0.00047325 = 0.33\%$.

2.6 PRACTICAL CONSIDERATIONS: ROUNDOFF

There is a feeling of success in the air. In most cases, mathematicians cannot only prove convergence by the ratio test; they can often prove convergence *to the right value* by Taylor's theorem (2.10) by putting a bound on the magnitude of the last derivative, over the whole jump, perhaps for any derivative, for any jump. For example, in the series (2.3) for sin x, every derivative is $\pm \sin x$ or $\pm \cos x$, which always lies between -1 and 1. Thus Taylor's theorem proves convergence. Indeed, we can make the enormous jump in Fig. 2.3, to $x = 13$ radians, merely by matching every derivative at $x = 0$, at the beginning of the jump, A, in Fig. 2.3. The mathematicians are right. We must not become blasé about Taylor's series. It is truly amazing.

Yet in practical terms, all is not sweet and beautiful. Perhaps in your maths courses, you have seldom if ever put numbers into a Taylor series—although it is easy nowadays, with your pocket calculator. What actually happens, if you put $x = 13$ in the series (2.3)? Directly from your calculator, $\sin 13 = 0.42017$, so we know the answer; but the seventh term in the series is 48639, so on the calculator we cannot avoid losing 5 decimals of accuracy!—if we want 5 places in the final answer, we must know the seventh term to 10 significant figures, for example. Or to get 10 decimal places, we should need a calculator which carries 15 significant figures, etc. It would be even worse with larger x. We have a case which is abnormally sensitive to roundoff errors. We do not disagree with the mathematicians' predictions, merely with their feasibility in practical terms. In the next chapter, we shall take another look at this phenomenon, but it is worth remembering that sometimes it is awkward to obtain derivatives to the desired accuracy. Also if the function is not sufficiently smooth, higher order derivatives might not even exist!

EXERCISES

2.1. Confirm Taylor's theorem for

(a) $e^1 = 1 + 1 + \dfrac{1}{2} + \dfrac{1}{6} + \dfrac{1}{24} e^{\xi}$

(b) $\cos (\theta) = 1 - \dfrac{\theta^2}{2!} + \dfrac{\theta^4}{4!} \cos (\xi\theta)$, for $\theta = 0.5$

Consider our formula for approximate termination.

2.2. Find the first four terms of the Taylor's expansion of $f(x) = \sin(x) + x + 1$ about $x = 0$ and $x = 1$.

2.3. Evaluate

$$\int_0^x e^{-(x^2/2)} \, dx$$

by expressing the integrand as a Taylor series, and integrating term by term, for (a) $x = 1$, (b) $x = 10$. Comment on the convergence.

2.4. The following represent a polynomial. Of what degree is the polynomial and what is the coefficient of the highest degree?

(x)	-3	-2	-1	0	1	2	3	4
$f(x)$	-765	-119	3	9	-5	-15	171	1153

2.5. For $f(x) = \cos x$, use central difference operators to find $f'(2)$ and $f''(2)$ with $h = 0.1$. Check the values you obtain with a smaller value of h and compare with values obtained by direct differentiation.

2.6. A projectile of mass 0.12 kg is fired vertically upwards with an initial velocity $v(0) = 8$ m/s. The projectile is slowed owing to the force of gravity $F_g = -mg$ and to air resistance $F_R = -kv^2$ where g is 9.81 m/s and $k = 0.002$ kg/m. The Differential Equation of upwards velocity is thus

$$m \frac{dv}{dt} = -mg - kv^2.$$

Use Taylor's method of order 4 (what does that mean?) and find:
(a) the velocity after 0.2, 0.4, 0.6, 0.8, and 1.0 s
(b) the time to the nearest 0.01 s when the projectile reaches its maximum height.

3

Roundoff, etc.—adventures with Pi

3.1 ORDINARY ROUNDOFF

This chapter is not entirely about roundoff. Perhaps roundoff is the most important thing in your mini-apprenticeship for the moment: but there are a great many things that people whose business is numbers will expect you to know about. Mostly they are fairly simple. 'Pi' is not simple, really. If you heard about it many years ago at school, maybe you had a wrong impression. Really, 'Pi' is a lame pretence that this chapter possesses a structure, which in fact it doesn't.

Roundoff. If your calculator carries only ten significant figures, and you do the following calculation:

$$\pi^3 = 31.006276$$
$$- \underline{31}$$
$$\underline{0.006276}$$

Too bad. In calculating $\pi^3 - 31$, you can't help losing 4 significant figures. Generally, in calculating $A + B = C$ the roundoff damage depends on the largest of A, B, and C. For example, if you are calculating $A + B + \cdots + Z$, where $A \ldots Z$ are all positive, but they get steadily smaller, $A \gg Z$, then it would probably be more accurate if you reversed the order, $Z + Y + X + \cdots + B + A$. You can seldom do this conveniently; but it's worth knowing, and it makes you think. Later in this chapter, we shall see that the precise way in which we do the arithmetic can be crucial.

In most respects, multiplication and division are simpler, where roundoff is concerned. When you calculate A/B, you may have to round A and B when you introduce them as data. Again, the process of division introduces roundoff, of order $\pm 10^{-10}$ as a proportionate error perhaps, $\pm 10^{-8}\%$. In

fact, it really deserves much more careful treatment, if you can think clearly
about statistics: the error would have a rectangular distribution, between
$-1/2$ and $+1/2$ in the last place. Even that is an over-simplification.

Actually, as most people know, your calculator performs ordinary
arithmetic, but computers use 2 instead of 10 (binary notation):

$$1101 = 1 \times 2^3 + 1 \times 2^2 + 0 \times 2^1 + 1 \times 2^0 = 13$$
$$0.101 = 1 \times \tfrac{1}{2} + 0 \times 2^{-2} + 1 \times 2^{-3} = \tfrac{5}{8}.$$

This means that every item of data must be translated, before it can be stored,
ready for arithmetic to begin. Because there is a limit to the number of 1's and
zeros (binary digits) which can be stored, the first roundoff damage is almost
always inflicted before the arithmetic even starts. But it is only perhaps once a
year or less that one needs to remember that most computers use binary
arithmetic.

Just about as often, you are expected to know that when computer people
are talking about roundoff, they are like lawyers: don't take the obvious
interpretation, the nearest tenth decimal for example. No; they have complicated algorithms, which make adjustments giving the right values on average,
as they should, but which give slightly greater standard deviation than we
would naively expect. Moral: don't be naive, but don't probe unless you have
to ... or you may waste a lot of time.

In Section 2.6, we questioned the wisdom of calculating sin 13 by Taylor's
series. A more spectacular example is sin $N\pi$, where N is an integer, and π is
defined to as many places as the computer takes (3.14159265358979323846).
This should give zero. By Taylor's series, using the greatest precision
available on our local computer:

N	sin $N\pi$	
1	-0.7 D-19	(N.B.—D denotes double precision exponential—
2	-0.6 D-17	more precise than E.)
4	-0.4 D-14	(12.6 radians, similar to Fig. 2.3)
7	-0.9 D-11	
11	0.6 D-5	
15	0.1 D1	
25	0.7 D12	
26	Overflow	

This deteriorates progressively, because you have to take more terms, and
because the largest term gets bigger each time.

3.2 SUPER ROUNDOFF

But you are almost certainly unaware of our next little game:

$$\pi_i = 4\pi_{i-2} - 3\pi_{i-1}. \tag{3.1}$$

We put $\pi_1 = \pi_2 = \pi$, to ten places; $\pi_3 = \pi$? Only approximately … program (3.1) for your pocket calculator and put π in storage locations 'A' and 'B'

> read 'B' (π_{i-2})
> multiply by 4
> read 'A' (π_{i-1})
> write to 'B' (to become π_{i-2} for next round)
> multiply by 3 and subtract (π_i)
> write to 'A' (to become π_{i-1} for next round)
> pause to look at it, and go back to the first instruction.

Let it run. Actually do this. (Or program it on your big computer.) In a minute or so your calculator will probably stop itself, because $\pi_n > 10^{100}$. Unless you see it happen, you won't believe it ….

This weird phenomenon is known as 'exponential roundoff'. You can explain it to yourself, if you record three successive values, while they are still in the hundreds or the thousands; on a simple calculator we obtained

$$\pi_{19} = 2590.23, \qquad \pi_{20} = -10345.20, \qquad \pi_{21} = 41396.52$$

$$\frac{\pi_{21}}{\pi_{20}} \cong \frac{\pi_{20}}{\pi_{19}} \cong -4.$$

So the error increases by a factor of 4 each time! Following this clue intuitively, let's assume that the error is compounded in each iteration such that

$$\pi_n = Ar^n$$

where r is a roundoff error and A is a constant.

$$Ar^i = 4Ar^{i-2} - 3Ar^{i-1}$$
$$r^2 + 3r - 4 = 0 = (r + 4)(r - 1)$$
$$r = 1 \text{ or } -4.$$

If $r = 1$, the value of π wouldn't change, which would be surprising since π is an irrational number. So $r = 4$ and the error increases rapidly: we experience super-roundoff!

An easier approach now. Suppose that because π is irrational, a roundoff error of ε occurs the first time the formula (3.1) is applied, but none thereafter:

$$\pi_1 = \pi$$

$$\pi_2 = \pi$$

$$\pi_3 = \pi + \varepsilon$$

$$\pi_4 = 4\pi_2 - 3\pi_3 = \pi - 3\varepsilon$$

$$\pi_5 = \pi + 13\varepsilon$$

$$\pi_6 = \pi - 51\varepsilon \text{ etc.}$$

Already the errors are increasing nearly 4:1 each time. (Of course, if there is no error the first time, with this assumption you would never see any roundoff. In that case place some trust in the computer.)

Hurrah, explained. Hurrah? If this is actually going to happen in real jobs, every other day, then numbers are decidedly *not* going to be fun? Give up?

No, don't give up. It doesn't happen at all frequently. And what you have just seen will be a good introduction to 'instability' in solving differential equations, which happens much more frequently, and which is disastrous only for people who are unaware of what is happening.

On the other hand, develop a healthy respect for roundoff! Don't get complacent—experience with roundoff generates a sort of respect. For example, try an experiment:

$$\pi_N = 3\pi_{N-1} + \pi_{N-3} - 3\pi_{N-2}, \qquad \pi_1 = \pi_2 = \pi_3 = \pi.$$

This gives $r = 1$ three times. Yet working overnight with our calculator, $\pi_{104} = 3.141642636$, $\pi_{1004} = 3.146592654$, $\pi_{12004} = 3.692870280$. The errors were much larger than we expected; they don't increase exponentially, but they aren't linear either: very roughly they are proportional to (number of iterations)^2. Be prepared for further shocks! Don't ever assume you *really* understand roundoff.

3.3 FRACTIONAL APPROXIMATIONS

Not roundoff, this time—these are deliberate approximations. This is a fun subject, which may prove unexpectedly useful to you one day.

For example, $\pi = 3.14159265358979\ldots \cong 22/7 \cong 355/113$. Have you seen $355/113 = 3.14159292\ldots$? How do you find good approximations like this? 'Continued fractions':

$$3.14159265358979\ldots = 3 + \cfrac{1}{7.06251\ldots}$$

$$= 3 + \cfrac{1}{7 + \cfrac{1}{15.996\ldots}}$$

$$= 3 + \cfrac{1}{7 + \cfrac{1}{15 + \cfrac{1}{1.0034\ldots}}}$$

$$= 3 + \cfrac{1}{7 + \cfrac{1}{15 + \cfrac{1}{1 + \cfrac{1}{292.635}}}}$$

Now,

$$3 + \cfrac{1}{7 + \cfrac{1}{15}} = \frac{333}{106} = 3.141509$$

$$3 + \cfrac{1}{7 + \cfrac{1}{15 + \cfrac{1}{1}}} = \frac{355}{113} = 3.14159292\ldots$$

$$3 + \cfrac{1}{7 + \cfrac{1}{15 + \cfrac{1}{1 + \cfrac{1}{293}}}} = \frac{104348}{33215} = 3.1415926539\ldots.$$

Each of these is the most accurate possible, using integers of that size or smaller. Easy, isn't it!

3.4 SLOWLY CONVERGING SERIES

Now for some fairly conventional mathematics:

$$\tan^{-1}x(\text{for } x = 1) = \frac{\pi}{4} \text{ radians} = \int_0^1 \frac{dx}{1 + x^2} = \int_0^1 1 - x^2 + x^4 - x^6 \ldots dx.$$

Integrating term by term:

$$= 1 - \frac{1}{3} + \frac{1}{5} - \frac{1}{7} + \frac{1}{9} \cdots \tag{3.2}$$

'Pi' again. But this is a peculiarly useless series—how many terms will you need for 6-decimal accuracy? 1000000? For 20-decimal accuracy? In fact, this series is used to compute π, but first we must transform it by 'Euler's formula'. To derive the theory we rewrite the difference table described in section 2.4, using a new notation; 'Calculus of Finite Differences' again:

$$u_1$$
$$\Delta u_1(=u_2 - u_1)$$
$$u_2 \qquad\qquad\qquad\qquad \Delta^2 u_1(=\Delta u_2 - \Delta u_1)$$
$$\Delta u_2(=u_3 - u_2)$$
$$u_3 \qquad\qquad\qquad\qquad \Delta^2 u_2(=\Delta u_3 - \Delta u_2)$$
$$\Delta u_3 \text{ etc.}$$
$$u_4$$

Let's define the two 'operators' for this table:

$$Eu_n = u_{n+1},$$

so that E still means 'move vertically down one line'.

$$\Delta u_n = u_{n+1} - u_n = E u_n - u_n = (E - 1)u_n$$

so that Δ means 'move diagonally down, and to the right'. Not exactly the same as 'δ'. Evidently

$$\Delta = E - 1$$

and by ordinary algebra,

$$E = 1 + \Delta$$

Let's digress. For example, $\Delta^n u = (E - 1)^n u$; thus $\Delta^2 u_1 = u_3 - 2u_2 + u_1$. Check that this is so. It may be useful; Δ^n is akin to the nth derivative, like δ^n. If we know that the given function is a cubic, for example, then the fourth derivative must be zero and it follows that

$$\Delta^4 u_i = (E - 1)^4 u_i$$

$$= u_{i+4} - 4u_{i+3} + 6u_{i+2} - 4u_{i+1} + u_i$$

$$= 0 = \delta^4(u_{i+2}).$$

This is occasionally useful. Check it.

A more interesting application now. Let's look again at the mean slope, in terms of the Taylor series (section 2.3)

$$\delta = E^{1/2} - E^{-1/2} \qquad \text{equation (2.9), p. 28}$$

Remember, E means move down one line vertically: that is, go from $f(x)$ to $f(x + h)$. Taylor's theorem?! From the pocket version, equation (2.4), we can write

$$\delta = \exp\left(\frac{1}{2} h \frac{d}{dx}\right) - \exp\left(-\frac{1}{2} h \frac{d}{dx}\right)$$

$$= 1 + \frac{1}{2} h \frac{d}{dx} + \frac{1}{8} h^2 \frac{d^2}{dx^2} \cdots$$

$$- \left(1 - \frac{1}{2} h \frac{d}{dx} + \frac{1}{8} h^2 \frac{d^2}{dx^2} \cdots\right) \text{from the exponential expansion}$$

$$= h \frac{d}{dx} + \frac{1}{24} h^3 \frac{d^3}{dx^3} + \frac{1}{1920} h^5 \frac{d^5}{dx^5} \cdots \qquad (3.3)$$

which confirms our previous experience with the difference table.

The mind-stretching techniques of 'Finite Difference Calculus' would express this in an even more sophisticated way:

$$\delta = 2 \sinh\left(\frac{1}{2} h \frac{d}{dx}\right)$$

so that:

$$h \frac{d}{dx} = 2 \sinh^{-1}\left(\tfrac{1}{2}\delta\right).$$

Expanding this by Taylor's series:

$$h \frac{d}{dx} = \delta - \frac{1}{24}\delta^3 - \frac{31}{69120}\delta^5 \cdots$$

Could we do the opposite—could we integrate by differencing in reverse, factoring by h, adding some small corrections, etc.? Wait and see ...

The calculus of finite differences will even help us to calculate π! Start with the standard formula (3.2), and tabulate the terms, ignoring the sign:

u	Δu	$\Delta^2 u$	$\Delta^3 u$	$\Delta^4 u$
1.0000				
	−0.6667			
0.3333		0.5334		
	−0.1333		−0.4572	
0.2000		0.0762		0.4063
	−0.0571		−0.0509	
0.1429		0.0253		
	−0.0318			
0.1111				

Now write the original series in 'operational' form

$$\frac{\pi}{4} = u_1 - u_2 + u_3 - u_4 \cdots$$

$$= (1 - E + E^2 - E^3 \ldots)u_1.$$

We recognise this series as the Taylor expansion of $1/1 + E$ about zero. Let's play games with this version:

$$\frac{\pi}{4} = \frac{1}{1 + E}u_1$$

$$= \frac{1}{1 + (1 + \Delta)}u_1$$

$$= \frac{1}{2}u_1 - \frac{1}{4}\Delta u_1 + \frac{1}{8}\Delta^2 u_1 \ldots \text{(expanding as a binomial series)} \quad (3.4)$$

This is the 'Euler Transformation' of the series, $u_1 + u_2 + u_3$ (more formally, the sum of the series),

$$S = \sum_{i=1}^{\infty} \frac{(-1)^{i-1}\Delta^{i-1}u_1}{2^i}$$

Substituting from the difference table for our series for $\pi/4$:

$$\frac{\pi}{4} = \frac{1}{2}(1) - \frac{1}{4}(-0.6667) + \frac{1}{8}(0.5334) - \frac{1}{16}(-0.4572) + \frac{1}{32}(0.4063)$$

$$= 0.7746$$

so $\pi = 3.0984$. Quite good: not brilliant. This 'accelerated' series for π takes the general form

$$\pi = 2\left\{1 + \frac{1}{3} + \frac{1.2}{3.5} + \frac{1.2.3}{3.5.7} \cdots\right\} \tag{3.5}$$

Check this out by taking our equation for $\pi/4$, multiplying up, and converting the decimals to fractions. So

$$\pi = 2\{1 + 0.33333 + 0.13333 + 0.05714 + 0.02540 + 0.01154 \cdots\}$$
$$= 3.12148$$

Better, we can assume that subsequent terms decay in the ratio 2:1.

$$2\{1 + 0.33333 + 0.13333 + 0.05714 + 0.02540 + 2 \times 0.01154\} = 3.14456$$

because

$$1 + \frac{1}{2} + \frac{1}{4} + \frac{1}{8} \cdots = 2.$$

3.5 COMPUTING PRACTICE

The obvious way to calculate the fourth term of (3.5) is as $T_4 = 6/105 = 0.05714\ldots$ but as we tried to warn you in the first section of this chapter, you must learn to ask questions: like, is this the cheapest way to compute (3.5)? Obviously not, in this case. For example, T_5 will be $T_4 \times 4/9$. If we compute T_{n+1} as $T_n n/(2n + 1)$, which we add into the series, then we have to do 2 multiplications, 1 addition, and 1 division for each extra term. Doing it the obvious way, we would have to do $(n - 2)$ multiplications and $(n - 1)$ divisions for the nth term. We should always avoid making naive assumptions, not only when calculating infinite series. Even computing a simple polynomial, like

$$1.4x^7 - 2.7x^6 + 4.1x^5 + 3.2x^4 + 1.9x^3 + 0.6x^2 + x + 5$$

we should proceed as follows:

$$((((((1.4x - 2.7)x + 4.1)x + 3.2)x + 1.9)x + 0.6)x + 1)x + 5. \tag{3.6}$$

This involves only 7 multiplications. Most students would write the Fortran expression

$$1.4*x**7 - 2.7*x**6 \cdots.$$

If this logs and then delogs, it is certainly worse. If it makes use of 7 in its binary form, it is somewhat better, but (3.6) would remain much better. However, whether you should use (3.6) on your calculator depends on the commands available, and the speed. Experiment a little.

One of the most pleasant things about numerical methods, is that you are constantly meeting ingenious, almost magic little money-saving tricks like this. You sometimes wonder how people invented them. We shall see one involving triple matrix products. Unfortunately, we shall not see the most magic, the Fast Fourier Transform, which reduces the arithmetic by orders of magnitude; it would be too difficult.

3.6 HIGH ACCURACY CALCULATIONS

With formula (3.5) the dedicated 'computer-buff' can calculate π to hundreds of places! Something you only read about in books.... The trick is *not* to use floating-point arithmetic. Writing the answer we expect in the form

$$\text{NPI} (60) = (3, 141, 592, 653, 589, 793, 238, \ldots)$$

that is, the number π will be expressed as an array, containing a succession of integers, each representing three decimal places of π. We need another array, to represent the present term, and we need another subroutine to accumulate this into NPI. We must also write two subroutines, to multiply such a number by an integer, and to divide by an integer, in the scale of 1000, following all the rules we learnt as children. Thus NTERM $= 2 = (2, 0, 0, 0\ldots)$ initially, for the first term; it is divided by 3 for the second term; then it is multiplied by 2 and divided by 5 for the third term; and so on. Each term tends to be half its predecessor, so for 180 decimals, we can expect about $180/\log_{10} 2 = 598$ terms. The computer time is probably acceptable.

The computer-buff tends to be a lonely, intelligent person whose only joy is to interact with the computer. It seems a pity if such people are turned off by a course in numerical analysis, when they have so much genuine enthusiasm to offer.

EXERCISES

3.1. Express $\sqrt{2}$ as an infinite continued fraction. For some obscure engineering reason, we want a gear ratio $\sqrt{2}:1$. The number of teeth in both gears must be between 100 and 200. Choose them.

3.2. How would you compute economically $\cos x = 1 - x^2/2! + x^4/4!\ldots$.

3.3. Deduce an Euler transformation, like series (3.5), for $\ln 2 = 1 - \frac{1}{2} + \frac{1}{3} - \frac{1}{4}\cdots$.

4

Introduction to graphics

4.1 ENGINEERS AND PICTURES

Mathematicians, computer folk, and even a few theoretical engineers, are very obviously surprised at the way that practical engineers like to communicate in pictures. We think physically about our problems; we even try to imagine that we ourselves form a part of the mechanisms, etc. When this kind of tactile imagery is difficult, we are ill at ease. So it is hardly surprising that for many years engineers have been willing to spend hundreds of dollars for a crucial picture, whether drawn by a human or by a computer—for example, of a very complex casting. It clarifies things. Increasingly, computers are helping us to draw pictures, by turning pictures into numbers, by manipulating the numbers, by storing the numbers, etc. It is an exciting field. Computer graphics opens up many new possibilities, and already a lot of computing money is spent with a good conscience. Traditionally it is not a part of 'numerical methods' courses, but perhaps it should be, for students in engineering.

Around 1960, people were already thinking about computer graphics, but their pictures were crude 'wire models' (the problem of 'hidden lines' was not given top priority) and when the pictures moved, cartoon-fashion, they staggered about in slow jerks—crude, dinosaurs, compared with modern video-games. Around 1970, people began to create 'synthetic photographs' by defining the greyness at any point on a surface in terms of its orientation. This was expensive. Nowadays the expensive operations are those that deal with shadows, reflections, and refraction inside transparent objects; the modern results cannot be distinguished from photographs—except by connoisseurs, who insist that the choice of lighting, etc., is not up to professional photographic standards!

4.2 PERSPECTIVE

This introduction to graphics approaches perspective in three gentle stages: perspective with the eye and the picture in standardised positions relative to the x, y, z axes; ditto, with generalised geometry; and finally, the problem of choosing the best orientation for the picture frame, given the eye position. This introduction includes a comprehensive review of vectors, which is useful in itself; vectors are a godsend in 3D thinking—they should be second nature to any practising engineer.

1. In Fig. 4.1 the point on the object, Q, is (x, y, z). To simplify the algebra, the picture frame is in displaced x, z coordinates, now known as (X, Z). Thus P, which you will plot, will be in coordinates X, Z. The viewing distance OF is d. By similar triangles, taking a plan view from above,

$$X = \frac{xd}{y}, \tag{4.1}$$

and projecting into the yz plane, i.e. taking a side elevation,

$$Z = \frac{zd}{y}. \tag{4.2}$$

2. In Fig. 4.2 the point Q on the object is given in global (x^1, y^1, z^1) axes as $\vec{Q}$. First, the origin is shifted to the eye, E, by subtracting, giving,

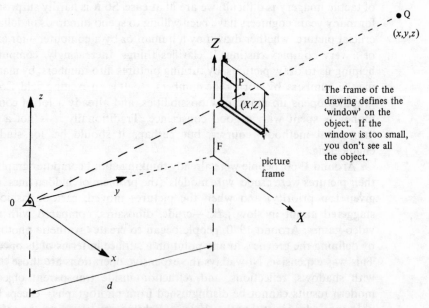

The frame of the drawing defines the 'window' on the object. If the window is too small, you don't see all the object.

Fig. 4.1   The ray OQ is a straight line, cutting the picture frame FXZ at P, which in picture coordinates is (X, Z).

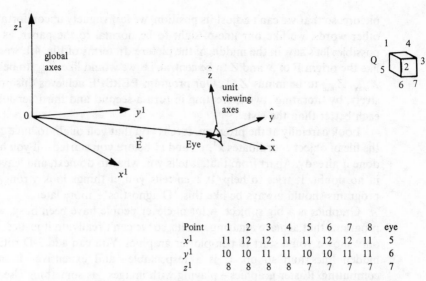

Point	1	2	3	4	5	6	7	8	eye
x^1	11	12	12	11	11	12	12	11	5
y^1	10	10	11	11	10	10	11	11	6
z^1	8	8	8	8	7	7	7	7	7

Fig. 4.2 In global (x^1, y^1, z^1) axes the eye position is $\vec{E}$ and the 'viewing axes' are (x, y, z) as before. The unit viewing axes are $\hat{x}, \hat{y}, \hat{z}$, in x^1, y^1, z^1 axes with $\hat{y}$ pointing 'towards' the object. The coordinates relate to the exercise at the end of the chapter.

$\vec{EQ} = \vec{Q} - \vec{E}$. The x-coordinate in the viewing axes is then given as the scalar product $\hat{x} \cdot (\vec{Q} - \vec{E})$, ($\hat{x}$ is in x^1, y^1, z^1 axes) because scalar multiplying any vector into the unit vector $\hat{x}$ implies multiplying it by the cosine between the vector and $\hat{x}$, i.e. resolving it in the x-direction. Putting y and z in place of x, we get all three coordinates (x, y, z) in the viewing axes, to substitute in formulae (1) and (2).

The unit 'viewing' vectors $(\hat{x}, \hat{y}, \hat{z})$ are stored as the three columns of the matrix $xyz(3, 3)$ in the program PERSPE (see Appendix), for convenience. Any matrix comprising three mutually orthogonal unit vectors is known as an 'orthogonal matrix'.

3. So far, we have ignored the only difficult problem: that of defining $\hat{x}, \hat{y}$, and $\hat{z}$ for the user; this is very necessary as the user may be an artist who is nearly innumerate. If $\hat{y}$ is known, we can create $\hat{x}$ in the horizontal plane, normal to $\hat{y}$, by vector multiplying $\hat{y}$ into a vertical unit vector.

$$\hat{x} = \hat{y} \times (0, 0, 1) \text{ normalised}$$

('normalised' means that we scale it to have unit length). Then $\hat{z}$ is easy: $\hat{z} = \hat{x} \times \hat{y}$. (vector product)

But how do we start, how do we choose $\hat{y}$? Largely by commonsense: We find the centroid of all the given points on the object. We then take the vector from the eye to the centroid, and normalise it; this gives the initial $\hat{y}$, the first estimate.

Finally, let's apply more commonsense. We normally hold a picture so that we are looking directly towards it. If somebody else is holding the picture, so that we can't adjust its position, we feel vaguely uncomfortable. In other words, we like our line-of-sight to be normal to the paper, as far as possible: let's say, in the middle of the picture. In terms of Fig. 4.1, we would like the origin F of X and Z to be central, i.e. we would like X_{min} to be minus X_{max}, Z_{min} to be minus Z_{max}. Our program PERSPE achieves this progressively, by 'iterating' twice, creating in turn a second and third version of y, each better than the last.

Look carefully at the program. It tells you that you ought to have created the file of object coordinates x^1, y^1, and z^1 before you started—if you haven't done it already. Apart from that, it tells you what to do next, and leaves you in no doubt. It tries to help. It even tells you, if things look wrong. Your programs should always be like this. 'Diagnostics'—more later.

Graphics is a big subject. A lot of clever people have been busy, writing programs that achieve amazing results, so we can't really do it justice. We've only shown you a start to pen-plotter graphics. You can add 3-D rotations, hidden lines and so on. It is a respectable—and expensive—branch of computing. Raster graphics—playing with images—is something else again. There is a good chance that you will all use graphics packages at some time.

4.3 DRAWING GRAPHS: THE HARD WORK APPROACH

Your computer could draw lovely smooth graphs too, if we could teach him how. He's champ at arithmetic, but he never went to high school, poor guy. You know what *you* mean by a 'smooth curve'. He doesn't. But he's getting there—you see.

For a modest start, go back to Taylor's idea—polynomials are a natural choice, for representing smooth functions. Say,

$$y = f(x) = a_0 + a_1 x + a_2 x^2 + a_3 x^3$$

or as far as we care to go. Suppose that we know the value y_1 at x_1, y_2 at $x_2 \ldots y_4$ at x_4. A glutton for hard work might write and solve the four equations for the four 'variables' $a_0 \ldots a_3$ to make the curve pass through the four points:

$$a_0 + a_1 x_1 + a_2 x_1^2 + a_3 x_1^3 = y_1, \qquad x_1 = 1, y_1 = 1 \text{ in Fig. 4.3}$$
$$a_0 + a_1 x_2 + a_2 x_2^2 + a_3 x_2^3 = y_2, \qquad x_2 = 2, y_2 = 2$$
$$a_0 + a_1 x_3 + a_2 x_3^2 + a_3 x_3^3 = y_3, \qquad x_3 = 3, y_3 = 1$$
$$a_0 + a_1 x_4 + a_2 x_4^2 + a_3 x_4^3 = y_4, \qquad x_4 = 4, y_4 = 3$$

So presumably a cubic would be okay. We shall not pursue this approach but there will be something similar in the next chapter.

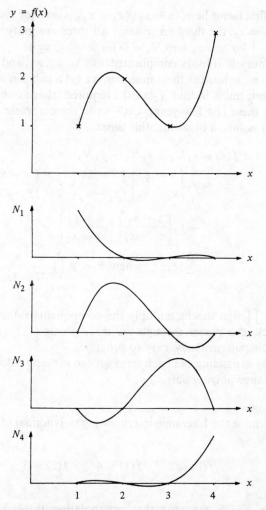

Fig. 4.3 The cubic through the four points marked x could be created as a linear combination of 'shape functions', $f(x) = N_1 + 2N_2 + N_3 + 3N_4$, where $N_4(x)$ for example is unity at x_4 and zero at x_1, x_2, and x_3.

4.4 A CLEVER TRICK

The idea of 'shape functions' outlined in Fig. 4.3 is attractive. (Shape functions are a concept which emerged from finite elements.) Who gets the prize for this idea?

Think about:

$$N_3(x) = \left\{\frac{x - x_1}{x_3 - x_1}\right\}\left\{\frac{x - x_2}{x_3 - x_2}\right\}\left\{\frac{x - x_4}{x_3 - x_4}\right\}$$

The first factor here, $(x - x_1)/(x_3 - x_1)$, is zero for $x = x_1$. The second is zero for $x = x_2$, the third for $x = x_4$. All three are unity, if we put $x = x_3$. Thus $N_3 = 1$ for $x = x_3$ and $N_3 = O$ for $x = x_1$, x_2 or x_4. Now try writing $N_1(x)$ for yourself. It looks complicated, but x_1, x_2, x_3, and x_4 are given constants: when we substitute these numbers, we get a cubic in x. So Lagrange wins our contest; this is indeed $N_3(x)$, the required 'shape function'. Pity, he has been dead these last few years. Let's write the complete function through *three* given points, a quadratic this time:

$$f(x) = y_1 N_1 + y_2 N_2 + y_3 N_3$$

$$= y_1 \left\{ \frac{x - x_2}{x_1 - x_2} \right\} \left\{ \frac{x - x_3}{x_1 - x_3} \right\} + y_2 \left\{ \frac{x - x_1}{x_2 - x_1} \right\} \left\{ \frac{x - x_3}{x_2 - x_3} \right\}$$

$$+ y_3 \left\{ \frac{x - x_1}{x_3 - x_1} \right\} \left\{ \frac{x - x_2}{x_3 - x_2} \right\} \dots \qquad (4.3)$$

$$= \sum_{i=1}^{n} F_i \qquad \text{where } F_i = y_i \prod_{\substack{j=1 \\ j \neq i}}^{n} \frac{x - x_j}{x_i - x_j}$$

(The $\prod$ sign means multiply the components rather than sum them ($\sum$).) Check, this passes through the three points (x_1, y_1), (x_2, y_2), and (x_3, y_3). Simple pattern really, easy to program.

By expanding out each term we can actually calculate a polynomial (the Lagrange polynomial).

Example
Determine the Lagrange interpolating polynomial of degree 3 or less for the following:

$$f(0) = 2 \qquad f(1) = 4 \qquad f(2) = 3 \qquad f(3) = 5$$

Solution
From above, we note that interpolating through n points will give a polynomial with a highest power of $n - 1$ (a degree of $n - 1$). So the polynomial through four points can be a cubic (or less if the coefficients of x^3 sum to zero).

Using equation 4.3,

$$f(x) = y_1 N_1 + y_2 N_2 + y_3 N_3 + y_4 N_4$$
$$= 2N_1 + 4N_2 + 3N_3 + 5N_4 \text{ (using the data)}$$

where

$$N_1 = \frac{(x - x_2)(x - x_3)(x - x_4)}{(x_1 - x_2)(x_1 - x_3)(x_1 - x_4)} \quad \text{etc.}$$

Here

$$x_1 = 0 \qquad x_2 = 1 \qquad x_3 = 2 \quad \text{and} \quad x_4 = 3.$$

So

$$N_1 = \frac{(x-1)(x-2)(x-3)}{(-1)(-2)(-3)} = -\frac{1}{6}x^3 + x^2 - \frac{11}{6}x + 1$$

$$N_2 = \frac{x(x-2)(x-3)}{(1)(-1)(-2)} = \frac{1}{2}x^3 - \frac{5}{2}x^2 + 3x$$

$$N_3 = \frac{x(x-1)(x-3)}{(2)(1)(-1)} = -\frac{1}{2}x^3 + 2x^2 - \frac{3}{2}x$$

$$N_4 = \frac{x(x-1)(x-2)}{(3)(2)(1)} = \frac{1}{6}x^3 - \frac{1}{2}x^2 + \frac{1}{3}x$$

$$\therefore \quad f(x) = x^3 - \tfrac{9}{2}x^2 + \tfrac{11}{2}x + 2$$

(For the coefficient of x^3, $2(-\frac{1}{6}) + 4(\frac{1}{2}) + 3(-\frac{1}{2}) + 5(\frac{1}{6}) = 1$ etc.)

4.5   A FORTRAN PROGRAM

Study the very small section of our program LAGRAN that would compute (4.3). The outer loop k introduces $y_k = y_1, y_2$, and y_3 in turn. The inner loop i tries to factor y_k three times, but fails because of the 'if' statement, which incidentally avoids dividing by zero.

But we insist that you study our programs in more depth. For example, if two of the given x values happened to be equal, then we divide by zero, despite the 'if' statement—disaster. 'Diagnostics' means any test that you apply to the data, before you allow computing to start, or sometimes during the calculation, to tell the user what mistake he has made. It saves time. Time is money. You must always have appropriate diagnostics in your programs.

Then again, we try to make the computer ask the user for precisely the data that are needed next, as he sits at the terminal. In the old days, with card input, you couldn't supply these 'prompts'! But nowadays, with terminals and micros, such 'user friendliness' is mandatory. Anybody who fails to conform will be—already is, in many offices—extremely unpopular in industry. Always do it—even if you will be the only one using the program! Try to get used to the idea now, while you are still learning.

Again, the listing of your programs should be self-explanatory. Put in plenty of 'comment' lines. When you are writing a long program, it will be easier even for you to find a particular statement, for example. But the important reason is that if the program is useful, somebody else will probably have to alter it later, when you are not around. You don't really want people to use foul language about you when you can't defend yourself! Of course, in the real world every program should be accompanied by a fully descriptive report and user guide.

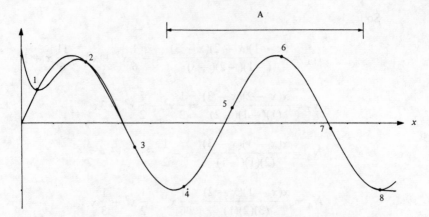

Fig. 4.4. Polynomial interpolation through points on a sine curve. In the region A the maximum error is only 0.01. But near and beyond the end-points the errors are intolerable. The behaviour is worse at the left-hand end.

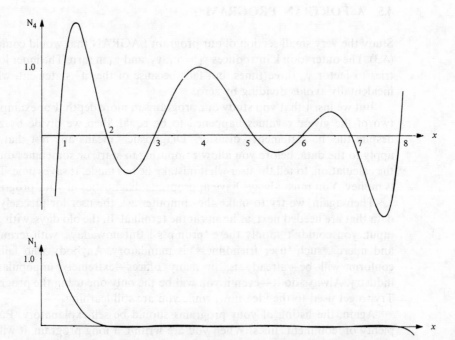

Fig. 4.5 The shape function N_4 (for a 'central' point) is quite wild at the ends, whereas N_1 (for an end point) is well-behaved in the middle. Because point 4 has a large value from the sine curve (Fig. 4.4), it makes a big contribution to the overall interpolation of the curve. Point 5, having a small value, does not upset the right-hand end as much. Perhaps a different set of points would have given a better approximation of the sine curve? Choose some and try it, or even wait 'til Chapter 7 and read all about it!

4.6 THE PROGRAM MISBEHAVES

The program enables us to fit a polynomial through any given points of any function (a cos function for the listing in the appendix), and then to compare the answers. This makes it a useful research tool. Figure 4.4 shows the results of a numerical experiment. We expect the errors to be very small near the centre of the range of given points, and the behaviour to become progressively wilder towards the ends. Here the results are quite good even at the right hand end. Beyond the ends it goes quite mad. Figure 4.5 gives a hint, but you will learn more about this wild behaviour in Chapter 7—it's not really as weird as you think.

Pity, though. Things seemed to be going so well. Never mind, we shall try again. It is possible to do much better near the ends, effectively, by using a springy beam to define the curve. See Chapter 13.

EXERCISES

4.1 With $\hat{x} = (0.6, -0.8, 0)$, $\hat{y} = (0.64, 0.48, 0.6)$ and $\hat{z} = (-0.48, -0.36, 0.8)$ and with the eye in the position shown, draw a perspective view of the cube in Fig. 4.2. Check that it makes geometric sense.

4.2 Suppose $f(x) = f(1)N_1(x) + f(2)N_2(x) + \cdots f(12)N_{12}(x)$ where N are Lagrange shape functions. Calculate
   (i) $N_{12}(6.5)$
   (ii) $N_2(11.7)$

4.3 For $f(x) = \sqrt{x}$, calculate values of $f(x)$ for $x_0 = 1.05$, $x_1 = 1.10$, $x_2 = 1.15$ and $x_3 = 1.20$. Estimate the value of $f(1.12)$ by (a) linear interpolation, (b) Lagrange interpolation, and compare with the actual value.

4.4 Evaluate the Lagrange polynomial of degree 3 or less which agrees with $f(1) = -2$; $f(2) = -2$; $f(4) = 4$; $f(5) = 10$.

4.5 Experiment with our program LAGRAN; change the positions of the points when interpolating cos x between say 0 and π. See what happens when you have a 'big' gap between the middle points. Try a different function also by changing the function (line 1) or writing a function subroutine.

4.6 The population of a city is given below. Use Lagrange interpolation to estimate the population in 1935 and 1968.

Year	1890	1900	1910	1920	1930
Population	67 210	75 555	83 567	91 722	120 405
Year	1940	1950	1960	1970	1980
Population	141 391	179 312	205 618	254 341	654 211

4.7 The viscosity of gasoline at four different temperatures is given below.
What is the *second*-order Lagrange polynomial which should be used to
interpolate for the viscosity at 85°C and what is that value?

$T(°C)$	$\mu(mPa \cdot s)$
0	0.71
50	0.43
100	0.29
150	0.22

4.8 Estimate the vapour pressure of propane at 12°C from the data below,
using Lagrange to find:
(a) 1st-order approximation
(b) 2nd-order approximation

$T(°C)$	$P(kPa)$
−100	3
−75	19
−50	75
−25	190
0	420
25	880
50	1800

5

Scattered points, and 'best' approximations

5.1 THE 'BEST' STRAIGHT LINE(S)

When we do an experiment, we record our measurements in a table. Then, teacher usually says we must plot the points, and draw a graph through them. All our readings are slightly wrong, everything misbehaves, so what is wanted is an 'average' line which then represents the truth, sort of. But don't complain. However many millions of dollars your equipment may cost in later life, nobody can pretend that the measurements will ever be *exact*. Whatever we do, we can't avoid introducing 'statistics' somewhere—that is, we have to make the best of things, admitting that the measurements are slightly wrong. What should our philosophy be? To minimise the biggest 'errors' from our curve? To average our results at one temperature, then at another and so on, then draw a smooth curve through the average values? To follow our instincts, to draw it by eye?

Probably the last is best, given the experience that you already have: but your computer has neither eyes nor instincts. This is why our computer-solution may seem artificial at first: following the standard procedure, we minimise the *sum* of the *squares* of the errors. Big error, big penalty—always positive. It works. Thus if we have just the four points $(1, 0)$, $(1, 1)$, $(2, 1)$, and

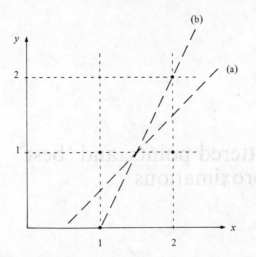

Fig. 5.1   Four given points: the problem is to put the most appropriate straight line through them. The choice, although obvious, depends on whether we regard the error as being in the y-direction (a), or the x-direction (b). (Rotate the book 90° clockwise.)

$(2, 2)$ as in Fig. 5.1, or (x_i, y_i), $i = 1$ to N in general, and we want to fit the straight line $y = a + bx$, our computer seeks to minimise

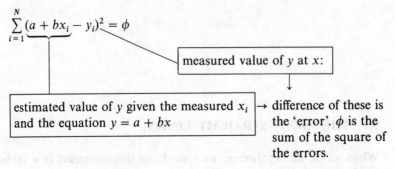

$$\sum_{i=1}^{N} (a + bx_i - y_i)^2 = \phi$$

measured value of y at x:

estimated value of y given the measured x_i and the equation $y = a + bx$ → difference of these is the 'error'. ϕ is the sum of the square of the errors.

Thus, apart from a factor of 2, we minimise ϕ with respect to both a and b in turn:

$$\frac{\partial \phi}{\partial a} = \sum_{1}^{N} (a + bx_i - y_i) = 0$$

$$\frac{\partial \phi}{\partial b} = \sum_{1}^{N} x_i(a + bx_i - y_i) = 0$$

Substituting x_i and y_i for the four points of Fig. 5.1 gives

$$4a + 6b = 4$$

$$6a + 10b = 7.$$

Thus $a = -\frac{1}{2}$, $b = 1$ as in the line (a) of Fig. 5.1, $y = x - \frac{1}{2}$. This is common sense: it minimises the greatest error, $\frac{1}{2}$; also it is the line through the mean of the two points at $x = 1$, and of those at $x = 2$. Nobody would argue.

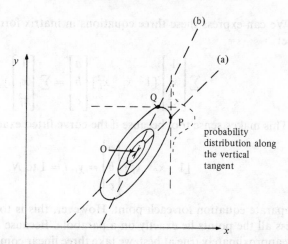

Fig. 5.2   A probability distribution of where the points might lie. The centroid of the distribution is at O. The ellipses are surfaces of equal probability density. The dotted curve shows the probability distribution along the tangent of one of the ellipses. The maximum density, which is also the mean along y, is at the point of tangency, P. So the line (a) as in Fig. 5.1 passes through P. Similarly (b) passes through Q. Thus we can argue that (b) is always steeper than (a).

However, if we interchange x and y, so that we seek the 'best' fit $x = a + by$, we find

$$4a + 4b = 6$$

$$4a + 6b = 7.$$

Thus $a = 1$, $b = \frac{1}{2}$, so $x = 1 + \frac{1}{2}y$ as in Fig. 5.1, line (b), or $y = 2x - 2$.

Don't be surprised. In statistics, the answers often depend strongly on the philosophy we adopt, as here. Without getting too deep, we can see roughly what is happening, in Fig. 5.2. This argument makes it clear, why line (b) is always steeper than (a). The two lines are identical if (and only if) all the given points lie on the same straight line!

5.2   OPTIMUM CURVES, SCATTERED POINTS

Let us now repeat the argument, with a quadratic, $y = a + bx + cx^2$; this may sometimes give a better representation of y:

$$\phi = \frac{1}{2} \sum (a + bx_i + cx_i^2 - y_i)^2.$$

For minimum ϕ

$$\partial \phi / \partial a = \sum (a + bx_i + cx_i^2 - y_i) = 0$$

$$\partial \phi / \partial b = \sum x_i(a + bx_i + cx_i^2 - y_i) = 0$$

$$\partial \phi / \partial c = \sum x_i^2(a + bx_i + cx_i^2 - y_i) = 0.$$

We can express these three equations in matrix form, which is easier to write:

$$\sum \begin{bmatrix} 1 \\ x_i \\ x_i^2 \end{bmatrix} \begin{bmatrix} 1 & x_i & x_i^2 \end{bmatrix} \begin{bmatrix} a \\ b \\ c \end{bmatrix} = \sum \begin{bmatrix} 1 \\ x_i \\ x_i^2 \end{bmatrix} y_i$$

This makes sense too, because if the curve fitted exactly, we should have

$$\begin{bmatrix} 1 & x_i & x_i^2 \end{bmatrix} \begin{bmatrix} a \\ b \\ c \end{bmatrix} = y_i, \, i = 1 \text{ to } N.$$

A separate equation for each point. However, this is too good to hope for, unless all the points lie exactly on a parabola. Because the N equations are only approximately true at best, we take three linear combinations, and make the mean equations true. We do our best, to solve all the equations exactly, but it is not possible unless we can make the errors zero; the smallest value that any sum of squares can take is zero!

Now, we apply our knowledge in a slightly different way. Suppose we have piles of different size aggregates in a gravel pit. A pile of coarse aggregate (relatively large size pieces), a pile of sand, and a pile of really fine particles (fines). For a concrete we want to blend proportions of these three aggregates to give an overall gradation close to the form

$$A = 100 \left(\frac{d}{D} \right)^{1/2}$$

where A is the percent of the aggregate passing a sieve size d (particle diameter) and D is the largest sieve size (largest particle diameter).

We decide to blend the three aggregates to give the least squares fit to the desired gradation. From sieve analysis we have:

Sieve size	14 mm	10 mm	5 mm	2.5 mm	1.25 mm	630 μm	315 μm	160 μm	80 μm
Aggregate	% Passsing								
C (coarse)	65.6	21.9	0.6	0.5	0.5	0.5	0.5	0.5	0.5
S (sand)	100	100	100	81.5	15.9	1.2	0.7	0.5	0.3
F (fines)	100	100	100	100	100	99.9	99.6	61.0	23.8

We calculate from the formula:

A (desired gradation)	100	86.5	59.8	42.3	29.9	21.2	15.0	10.7	7.6

Assume the final mix will contain proportions of x, y, and z of each of the coarse, sand, and fine aggregates.

We note that $x + y + z = 1$ (the three parts must make the whole), so we can eliminate one variable.

The difference from the desired gradation, at the ith sieve, is thus

$$D_i = A_i - C_i x - S_i y - F_i z$$
$$= (A_i - F_i) - x(C_i - F_i) - y(S_i - F_i).$$

There is a value of D for each sieve. The square of these differences, or 'error'

$$\phi = \sum_{i=1}^{n} D_i^2$$

for minimum ϕ

$$\frac{\partial \phi}{\partial x} = 0 = \sum_{i=1}^{n} [(A_i - F_i)(C_i - F_i) - x(C_i - F_i)^2 - y(S_i - F_i)(C_i - F_i)]$$

$$\frac{\partial \phi}{\partial y} = 0 = \sum_{i=1}^{n} [(A_i - F_i)(S_i - F_i) - x(C_i - F_i)(S_i - F_i) - y(S_i - F_i)^2]$$

$\therefore$ In matrix form, $\begin{bmatrix} a & b \\ b & c \end{bmatrix} \begin{bmatrix} x \\ y \end{bmatrix} = \begin{bmatrix} d \\ e \end{bmatrix}$

where

$$a = \sum_{i=1}^{n} (C_i - F_i)^2 \qquad d = \sum_{i=1}^{n} (A_i - F_i)(C_i - F_i)$$

$$b = \sum_{i=1}^{n} (C_i - F_i)(S_i - F_i) \qquad e = \sum_{i=1}^{n} (A_i - F_i)(S_i - F_i)$$

$$c = \sum_{i=1}^{n} (S_i - F_i)^2.$$

With the data given, we obtain:

$$a = 60868.14 \qquad d = 37549.78$$
$$b = 34028.27 \qquad e = 26521.34$$
$$c = 31150.46$$

hence

$$x = 0.362$$
$$y = 0.456$$
$$\therefore \quad z = 0.182$$

Our mix should have: 36.2% of coarse aggregate, 45.6% of the sand, and 18.2% of the fines.

The results of our efforts are shown in Figure 5.3. Note how the blended gradation swaps from side to side of the desired. We are ready for the next problem: you can probably guess it.

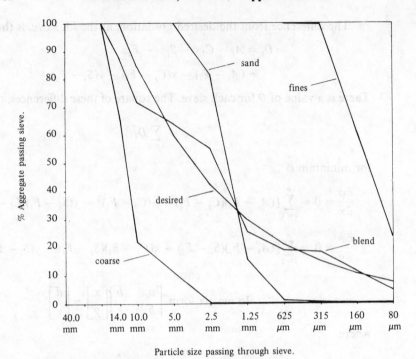

Particle size passing through sieve.

Fig. 5.3 Plots of the gradation of the three starting aggregates, with those of the desired and resulting blends.

5.3 APPROXIMATING A FUNCTION

We are given a complicated function. We would prefer to replace it by a simpler function, which is sufficiently accurate, and which is cheaper to calculate. Your calculator doesn't use Taylor's series for $\sin \theta$—that would take for ever!

Example
Consider $y = e^x$. We want to find the best fit $y = a + bx + cx^2$ to this function, between $x = 0$ and $x = 1$.

Solution
Treat this as another 'scattered point' problem, with an infinity of points, weighted dx for the region between x and $x + dx$, so that $\int$ replaces $\sum$:

$$\int_0^1 \begin{Bmatrix} 1 \\ x \\ x^2 \end{Bmatrix} [1 \ x \ x^2] \begin{Bmatrix} a \\ b \\ c \end{Bmatrix} dx = \int_0^1 \begin{Bmatrix} 1 \\ x \\ x^2 \end{Bmatrix} e^x \, dx.$$

(Note that the matrix expressions which form the integrands are the same as in the previous section.)

$$= \int_0^1 \begin{bmatrix} 1 & x & x^2 \\ x & x^2 & x^3 \\ x^2 & x^3 & x^4 \end{bmatrix} \begin{Bmatrix} a \\ b \\ c \end{Bmatrix} dx = \begin{Bmatrix} e^x \\ xe^x - e^x \\ x^2e^x - 2xe^x + 2e^x \end{Bmatrix} \Bigg|_0^1$$

$$= \begin{Bmatrix} e - 1 \\ 1 \\ e - 2 \end{Bmatrix}$$

$$\begin{bmatrix} 1 & \frac{1}{2} & \frac{1}{3} \\ \frac{1}{2} & \frac{1}{3} & \frac{1}{4} \\ \frac{1}{3} & \frac{1}{4} & \frac{1}{5} \end{bmatrix} \begin{Bmatrix} a \\ b \\ c \end{Bmatrix} = \begin{Bmatrix} 1.71828 \\ 1 \\ 0.71828 \end{Bmatrix}$$

Solving these equations gives

$$y = 1.0129913 + 0.8511251x + 0.8391840x^2.$$

The peak errors are 0.012991 at $x = 0$

$$0.014981 \text{ at } x = 1.$$

The error is zero at

$$x = 0.11623$$

$$x = 0.50891$$

$$x = 0.89089.$$

Okay, it crosses near the middle of the range, and near the ends; the maximum errors are at the two ends, and they are nearly equal. Hardly surprising. But these observations will confirm an approximate theory, in Chapter 7. A very practical consequence.

5.4   GENERAL ADVICE

If you have six points, and you know there are errors in the measurements, don't try to be clever, and fit a quintic! It can be done, but it is counter-productive; you are trying to extract too much information from your limited data. 'Economy' in the formula you are fitting is wise not only in the monetary sense, but also by common sense. Don't be pretentious. Much better, observe that the points nearly fit a log curve, or $y \cong a + b/x$ for example. If you are estimating only two constants, a and b, six points should be marginally adequate. It sometimes helps, to judge what happens at zero, or infinity, and think whether what you are doing could lead to absurd conclusions. For example, the infinity in $a + b/x$ may make you think again.

We have two programs for this chapter; LEASQ and CURFIT. LEASQ will fit a least squares polynomial through your points. Because we have the number of terms in the polynomial (nterm) as 4, the program will fit a cubic as is. CURFIT will approximate a cosine curve with a cubic over a range of your choice. You can change the order of the polynomials or the function if you wish, but do try a few experiments.

EXERCISES

5.1. Given the following data, find the best straight line:

x	1.5	2.4	3.6	4.2
y	10.6	12.9	17.8	21.5

5.2. Repeat the worked example with

Sieve size	14 mm	10 mm	5 mm	2.5 mm	1.25 mm
Aggregate	% Passing				
Coarse	100	69.2	15.2	0.5	0.4
Sand	100	99.6	99.1	87.4	78.7
Fine	100	100	100	100	100

Sieve size	630 μm	315 μm	160 μm	80 μm
Aggregate	% Passing			
Coarse	0.4	0.3	0.2	0.2
Sand	68.6	26.7	4.2	4.0
Fine	100	100	100	100

5.3. Find a least squares quadratic fit to the following data:

x	0.8	2.1	2.7	4.5	5.5	6.3	7.7	8.1	8.9	9.9	11.1	12.5
$f(x)$	17.1	15.5	14.0	19.0	24.6	28.0	41.0	47.3	55.5	70.3	89.2	114.2

5.4. Find a quadratic fit $a + bx + cx^2$ to the function $1/x$ between $x = 1$ and $x = 2$. Discover the points where it is exact. (Note: it will save a little time to note that the inverse of

$$\int_1^2 \begin{vmatrix} 1 & x & x^2 \\ x & x^2 & x^3 \\ x^2 & x^3 & x^4 \end{vmatrix} dx = \begin{vmatrix} 1 & \frac{3}{2} & \frac{7}{3} \\ \frac{3}{2} & \frac{7}{3} & \frac{15}{4} \\ \frac{7}{3} & \frac{15}{4} & \frac{31}{5} \end{vmatrix} \text{ is } \begin{vmatrix} 873 & -1188 & 390 \\ -1188 & 1632 & -540 \\ 390 & -540 & 180 \end{vmatrix}$$

5.5. The water level in a particular sea is determined by a tide whose period is about 12 hours and which may have the form

$$H(t) = h_0 + a_1 \sin \frac{2\pi t}{12} + a_2 \cos \frac{2\pi t}{12}$$

where t is in hours.

Given the following crude measurements estimate $H(t)$ using least squares.

t(hrs)	0	2	4	6	8	10
$H(t)$	1.0	1.6	1.4	0.6	0.2	0.8

5.6. Estimate the height above the sea level for the three points A, B, and C in the least squares sense, where the difference in altitude is measured crudely as shown, and D, E, and F lie at sea level.

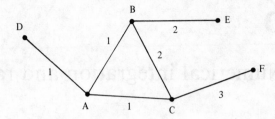

6

Numerical integration and ramifications

6.1 SUCCESS AT LAST

In our study of numerical techniques, maybe you feel little cause for optimism so far? Hang on, this is it. Remember the trouble you had, in doing integrals like $\int (a + b \cos \theta)^{-1} \, d\theta$? That was for the birds.... In real life, such tasks are usually done numerically, or by looking in a book of standard integrals. In the second case, it is always best to check the formulae numerically. To err is all too human.

6.2 VERY BASIC IDEAS

The trouble with studying numerical integration, sometimes known as 'quadrature', is that it tends to degenerate into merely learning a long, boring, list of more or less useful formulae, without any guiding principles, and without even learning to discriminate amongst the various rules. That being so, and we appreciate the danger, it seems best to introduce the subject in much the same way that you first encountered integration—see Fig. 6.1.

Expressing (a) as a formula gives

$$\int_{x_1}^{x_N} f(x) \, dx \cong (x_2 - x_1)f(x_1) + (x_3 - x_2)f(x_2) + \cdots$$

or, if all the panels are of the same width h,

$$\int_{x_1}^{x_N} f(x) \, dx \cong h\{f(x_1) + f(x_1 + h) + \cdots\}. \tag{6.1a}$$

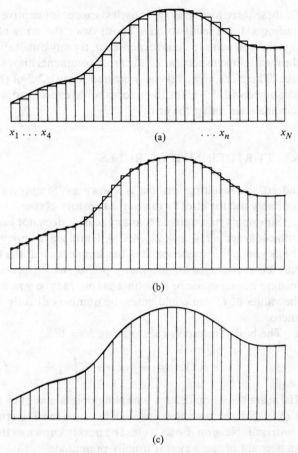

Fig. 6.1   (a) The model taken to introduce integration in mathematics. The value of the function $f(x)$ is assumed constant over any interval x_n to x_{n+1}, e.g.: x_1 to x_2, or from x to $x + dx$. The value of the integral is obviously too low for a function that persistently increases, but it tends to the correct answer in the limit, as dx tends to zero, i.e. as N becomes large. (b) If the panels do not tend to zero, a better assumption is that the mean value in every panel is the value halfway along that panel—which is reminiscent of the assumption in Chapter 2, that the 'mean slope' equals the 'midway slope'. (c) Finally, an assumption commonly applied is that the function varies linearly within each panel. This gives rise to the 'Trapezoidal Rule', because each panel consists of a trapezium, defined by the values of the function at the two ends. See again Fig. 6.3.

Expressing (b) in the same way:

$$\int_{x_1}^{x_1 + Nh} f(x)\, dx = h\left\{ f\left(x_1 + \frac{h}{2}\right) + f\left(x_1 + \frac{3h}{2}\right) + \cdots + f\left(x_1 + \frac{\overline{2N-1}h}{2}\right) \right\}.$$

(6.1b)

However, (c) is more untidy:

$$\int_{x_1}^{x_1 + Nh} f(x)\, dx = \frac{h}{2}\left\{ f(x_1) + 2f(x_1 + h) + 2f(x_1 + 2h) + \cdots \right.$$
$$\left. + 2f(x_1 + \overline{N-1}h) + f(x_1 + Nh) \right\}.$$

(6.1c)

Of these three formulae, (a) is useless except for captive maths students; (b) is quite good; (c) tends to have about twice the error of (b), and of opposite sign—if you want to do an experiment, try any quadratic, with $N = 1$; as with the mean slope in section 2.3 the error comments above turn out to be exactly true. Therefore, if we take a 'weighted mean', 2/3 of (b), the more accurate, combined with 1/3 of (c), the errors should cancel and we should have a better formula than either (b) or (c).

6.3 FURTHER SIMPLE RULES

Indeed, the resulting formula is known as 'Simpson's Rule', and you have probably used it already in your laboratory classes.

Simpson's rule, applied to many panels, does not justify the complication, with coefficients $1, 4, 2, 4, 2 \ldots 4, 2, 4, 1$, and we believe you will grow to prefer (b) as you gain experience. It is the simplest of all the formulae, so that you may be encouraged to use many points, which probably helps if you are reading from a crude or indistinct graph. Maybe you wouldn't even tabulate the values of f; you could enter the numbers directly on your calculator or micro.

The basic Simpson's rule, putting $N = 1$, is:

$$\int_{x_1}^{x_1 + h} f(x)\, dx = \frac{h}{6}\left[f(x_1) + 4f\left(x_1 + \frac{h}{2}\right) + f(x_1 + h)\right].$$

There are three equidistant 'sampling points' and they are multiplied by 1, 4, 1. (We shall see later how this is all that you need to remember.) It is the first nontrivial 'Newton–Cotes' rule. The next is known as the 'three-eighths' rule, on account of the way it is usually presented:

$$\int_{x_1}^{x_1 + 3h} f(x)\, dx = \frac{3h}{8}\left[f(x_1) + 3f(x_1 + h) + 3f(x_1 + 2h) + f(x_1 + 3h)\right]$$

but it is no more competitive than Simpson's rule. Incidentally the coefficients 1, 3, 3, 1 have nothing to do with $(1 + x)^3$. Our first recommendable rule is the next in the family:

$$\int_{x_1}^{x_1 + 4h} f(x)\, dx = \frac{4h}{90}\left[7f(x_1) + 32f(x_1 + h) + 12f(x_1 + 2h)\right.$$

$$\left. + 32f(x_1 + 3h) + 7f(x_1 + 4h)\right]. \quad (6.2)$$

The $7 - 32 - 12 - 32 - 7$ rule is worth remembering. Easy: (7, 32, 12, 32, 7) is all you have to know. Because $7 + 32 + 12 + 32 + 7 = 90$, we first divide by 90 to get the mean. Look at Fig. 6.2. Now we multiply by $4h$, the interval of integration.

'Weighted Means'—see Fig. 6.2. Suppose for example that 15% of families have single offspring, 40% have 2, 30% have 3, 10% have 4, and 5% have 5 children. Then the average family size is $0.15 \times 1 + 0.4 \times 2 + 0.3 \times 3 + 0.1 \times 4 + 0.05 \times 5 = 2.5$. Of course, the 'weights' 0.15, 0.4, 0.3, 0.1, 0.05

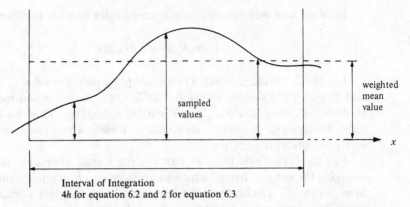

Fig. 6.2   The area under the curve equals the area of the equivalent rectangle. The height of the rectangle is a 'weighted mean' of the sampled values, i.e. $\sum$ (weighting factor) x (value) where $\sum$ (weighting factor) $= 1$. Otherwise the integration formula would fail for $f(x) = $ constant.

add up to 1 exactly—that is the criterion for a 'weighted mean'; otherwise we might call it just a 'linear combination'. We've met another weighted mean already—interpolation—although it wasn't worth mentioning at the time; why does $N_1 + N_2 + \cdots = 1$ (section 4.4) for any x? (Not at all obvious algebraically!) If it were not so, and $f(x) = 1$ for all x, i.e. for all the given points; then if we didn't get the answer $1 = N_1 + N_2 \ldots$ when we interpolated for any x, it would have to be wrong somewhere.... Subtle. But you can't argue!

In doing numerical integration, we calculate a genuine mean value—the average value of the function over the interval of integration; the sampling points in (6.2) are 0%, 25%, 50%, 75%, and 100% of the distance along the interval of integration. You will always find this; numerical integration is a matter of taking a 'weighted mean' of carefully chosen 'sampled function values', and then multiplying the mean by the 'interval of integration', to find the area of the rectangle in Fig. 6.2. If you always keep this basic idea in mind, you will avoid making the silliest mistakes.

6.4   GAUSS' SUPER-RULES

Look now at the Gauss 3-point rule:

$$\int_{-1}^{1} f(\xi)\, d\xi = \tfrac{5}{9}f(-\sqrt{0.6}) + \tfrac{8}{9}f(0) + \tfrac{5}{9}f(\sqrt{0.6}). \tag{6.3}$$

New idea. In this book we adopt the convention that a variable which goes from -1 to $+1$ is called ξ instead of x. Formula (6.3) optimises to great advantage where we choose to 'sample' the function, namely 11.3%, 50%, and 88.7% of the distance along the interval of integration. Note too, Gauss multiplies the sample values by 5/18, 4/9, and 5/18, which add up to 1. Then he multiplies the mean value by the interval of integration, 2, as in Fig. 6.2.

Let's see how well our two prizewinning rules actually perform:

$$\int_0^\pi \sin\theta \, d\theta = 2 \text{ exactly.}$$

For the 7, 32, 12, 32, 7 rule (1) the sampling points are $\theta = 0, \frac{1}{4}\pi, \frac{1}{2}\pi, \frac{3}{4}\pi$ and π, giving the sample values $0, 1/\sqrt{2}, 1, 1/\sqrt{2}$, and 0 and mean value $(7 \times 0 + 32/\sqrt{2} + 12 + 32/\sqrt{2} + 7 \times 0)/90 = 0.63616$. This looks about right. The integral $(0.63616 \times \pi)$ comes to 1.9986, which you must agree is very good under the circumstances.

For the Gauss rule (6.3) we can use the formal approach; this means changing the variable from θ which varies from a to b, say, to ξ, which varies from -1 to $+1$. The centre of the interval is $\frac{1}{2}(a+b)$, which corresponds to $\xi = 0$. The range is $b - a$, which corresponds to ξ varying from -1 to $+1$, a range of 2. Therefore we put $\theta = \frac{1}{2}(a+b) + \frac{1}{2}\xi(b-a)$, so that for $\xi = -1$, $\theta = a$ and for $\xi = 1, \theta = b$. We 'map' ξ onto θ for our integral. Putting $a = 0$, $b = \pi$, we have $\theta = \frac{1}{2}\pi(1 + \xi)$. Also $d\theta = \frac{1}{2}(b - a)\,d\xi, = \frac{1}{2}\pi \, d\xi$ here, so that transforming the integral,

$$\int_a^b f(\theta) \, d\theta = \int_{-1}^1 f\left(\frac{a+b}{2} + \xi \frac{b-a}{2}\right) \frac{b-a}{2} \, d\xi \qquad (6.4)$$

$$= \int_{-1}^1 f\left\{\frac{1}{2}\pi(1 + \xi)\right\} \frac{\pi}{2} \, d\xi.$$

Thus the sampling points are $\theta_i = \frac{1}{2}\pi(1 + \xi_i)$, $\xi_i = -\sqrt{0.6}, 0$ and $\sqrt{0.6}$, giving the values $\sin\theta_i = 0.34671, 1$, and 0.34671, so the Gauss rule becomes

$$\left(\frac{5}{9} 0.34671 + \frac{8}{9} + \frac{5}{9} 0.34671\right) \frac{\pi}{2} = 2.0014.$$

Thus we have the same error (actually a little less) for the cost of three sampling points instead of five. Knowledgeable people almost always prefer Gauss rules in computer programs! It may not be easier, but it's certainly cheaper, especially if f is complicated.

6.5 DERIVING INTEGRATION RULES

The most direct way to calculate the weighting factors is to use Lagrange's shape functions (section 4.4): for example, in Simpson's rule, we assume that

$$\int_1^3 f(x) \, dx = Af(1) + Bf(2) + Cf(3).$$

To find A, we choose for $f(x)$ the function that is 1 for $x = 1$ and zero for $x = 2$ or 3, so that $A = \int f(x) \, dx$ exactly; from Lagrange, section 4.4. The $f(x)$ we want is:

$$f(x) = \frac{x - x_2}{x_1 - x_2} \frac{x - x_3}{x_1 - x_3} = \frac{(x-2)(x-3)}{2}.$$

So, to find A,

$$\int_1^3 f(x)\,dx = A = \frac{1}{2}\int_1^3 (x^2 - 5x + 6)\,dx$$

$$= \frac{1}{2}\left[\frac{x^3}{3} - \frac{5x^2}{2} + 6x\right]\Bigg|_1^3$$

$$= \frac{1}{3}.$$

Because the shape functions are quadratics in x, it seems a fair assumption that Simpson's rule is exact for any quadratic. Indeed, a more formal way to calculate A, B, and C would be to insist that the formula gives the right answers for $f(x) = 1$, x, and x^2 in turn, and thus for any linear combination of these, i.e. any quadratic. Three points just define a quadratic interpolation (section 4.4) so it is highly reasonable to integrate that interpolation function. But surprise, Simpson is exact for $f(x) = x^3$ too! By shifting the origin, you can easily see this:

$$\int_{-1}^1 f(x)\,dx = 2[\tfrac{1}{6}f(-1) + \tfrac{2}{3}f(0) + \tfrac{1}{6}f(1)].$$

It is not immediately obvious, but ... by doing the integral, and entering $f(x)$, $f(x) = x^3$ gives zero on both sides. In the same way,

$$\int_{-2}^2 f(x)\,dx = \tfrac{4}{90}\left[7f(-2) + 32f(-1) + 12f(0) + 32f(1) + 7f(2)\right]$$

is exact, not only for the Lagrange shape functions, which would be quartics, but for x^5 too. Terrific.

The Gauss 3-point rule is comparable, judging from its performance. Let us feign ignorance of this rule and postulate

$$\int_{-1}^1 f(\xi)\,d\xi = H_1 f(\xi_1) + H_2 f(\xi_2) + H_3 f(\xi_3).$$

Suppose now that the rule is accurate for $f(\xi) = 1, \xi, \xi^2, \xi^3 \dots$ as far as we can manage:

$$\int_{-1}^1 1\,d\xi = 2 = H_1 + H_2 + H_3$$

$$\int_{-1}^1 \xi\,d\xi = 0 = \xi_1 H_1 + \xi_2 H_2 + \xi_3 H_3$$

$$\int_{-1}^1 \xi^2\,d\xi = \tfrac{2}{3} = \xi_1^2 H_1 + \xi_2^2 H_2 + \xi_3^2 H_3$$

$$\int_{-1}^1 \xi^3\,d\xi = 0 = \xi_1^3 H_1 + \xi_2^3 H_2 + \xi_3^3 H_3$$

$$\int_{-1}^1 \xi^4\,d\xi = \tfrac{2}{5} = \xi_1^4 H_1 + \xi_2^4 H_2 + \xi_3^4 H_3$$

$$\int_{-1}^1 \xi^5\,d\xi = 0 = \xi_1^5 H_1 + \xi_2^5 H_2 + \xi_3^5 H_3.$$

Six equations, six unknowns. Okay, but how do we solve them?

Actually, it isn't too hard to find ξ_1, ξ_2, and ξ_3. Let's consider

$$f(\xi) = (\xi_1 - \xi)(\xi_2 - \xi)(\xi_3 - \xi)$$
$$= a + b\xi + c\xi^2 - \xi^3 \text{ (say).}$$

Given a, b, and c, we could easily find the roots (our sampling points) ξ_1, ξ_2, and ξ_3 of the cubic $f(\xi)$. How to find a, b, and c? Observe that $f(\xi) = 0$ at each sampling point, so we expect the integral of $f(\xi)$ as given by the Gauss rule to be zero:

$$\int_{-1}^{1} f(\xi)\, d\xi = \int_{-1}^{1} (a + b\xi + c\xi^2 - \xi^3)\, d\xi$$
$$= 2a + \tfrac{2}{3}c = 0 \dots \tag{6.5a}$$

Further, because we expect the rule to be exact up to ξ^5, we should have

$$\int_{-1}^{1} \xi f(\xi)\, d\xi \text{ which also gives zero at the sampling points}$$

$$= \int_{-1}^{1} (a\xi + b\xi^2 + c\xi^3 - \xi^4)\, d\xi$$
$$= \tfrac{2}{3}b - \tfrac{2}{5} = 0. \tag{6.5b}$$

Similarly

$$\int_{-1}^{1} \xi^2 f(\xi)\, d\xi = \int_{-1}^{1} (a\xi^2 + b\xi^3 + c\xi^4 - \xi^5)\, d\xi$$
$$= \tfrac{2}{3}a + \tfrac{2}{5}c = 0. \tag{6.5c}$$

This is as far as we can go. Convenient: three equations, just enough to define $a = c = 0$, $b = 3/5$, so that

$$f(\xi) = \tfrac{3}{5}\xi - \xi^3 \tag{6.6}$$

giving roots of $\xi_1 = -\sqrt{(0.6)}$ $\xi_2 = 0$ $\xi_3 = \sqrt{(0.6)}$ which are the correct sampling points.

Now let's find H_1, H_2, and H_3: H_2 for example. Lagrange's 'shape function',

$$g(\xi) = \frac{(\xi - \xi_1)(\xi - \xi_3)}{(\xi_2 - \xi_1)(\xi_2 - \xi_3)} = \frac{\xi^2 - \tfrac{3}{5}}{-\tfrac{3}{5}}$$

is unity for $\xi = \xi_2$, zero for $\xi = \xi_1$ or $\xi = \xi_3$. Thus

$$\int_{-1}^{1} g(\xi)\, d\xi = H_1 g(\xi_1) + H_2 g(\xi_2) + H_3 g(\xi_3)$$

$$= H_2$$

$$= \int_{-1}^{1} \frac{\xi^2 - \tfrac{3}{5}}{-\tfrac{3}{5}}\, d\xi = \frac{8}{9}.$$

Okay. Gauss rules exist for any number of sampling points; they all give positive H_i, and the sampling points $\xi_1 \ldots \xi_n$ are always inside the interval of integration; that is, $-1 < \xi_i < 1$. Worth remembering but difficult to prove. Also, you must remember a simple formula for the polynomial giving $\xi_1 \cdots \xi_n$ as roots:

$$P_n(\xi) = \frac{1}{2^n n!} \frac{d^n}{d\xi^n} (\xi^2 - 1)^n.$$

With $n = 3$, this gives the cubic (6.6), apart from a factor which would make no difference to ξ_1, ξ_2 and ξ_3. $P_n(\xi)$ is known as the nth degree 'Legendre polynomial', and it is very important for people doing advanced numerical work, not just in numerical integration.

Gauss integration is a success story. Very extensive tables exist, not only for the family of basic Gauss rules (see, for example, our Table 6.3 at the end of this chapter), but for many, many variants. They behave well. They are not temperamental. This should be one of the happiest moments in your numerical methods course.

6.6   ROMBERG INTEGRATION

We now come to a topic† which is generally interesting, but in our opinion you are unlikely to use the idea for integration. Regard it as a teaching vehicle: 'Orders of Convergence'. First we must reintroduce the crudest and least effective rule known, the 'Trapezoidal Rule':

$$\int_0^{Nh} f(x)\, dx = h[\tfrac{1}{2}f(0) + f(h) + f(2h) + \cdots + f(\overline{N-1}h) + \tfrac{1}{2}f(Nh)].$$

We can demonstrate how Romberg integration works, with a simple example,

$$\int_0^8 x^4\, dx = \left[\frac{x^5}{5}\right]\Bigg|_0^8 = 6553.6.$$

We compute successively better approximations by the Trapezoidal rule:

$$A_8\left(\text{with } \begin{array}{l} h = 8 \\ N = 1 \end{array}\right) = 8[\tfrac{1}{2}0^4 + \tfrac{1}{2}8^4] = 16384$$

$$A_4\left(\text{with } \begin{array}{l} h = 4 \\ N = 2 \end{array}\right) = 4[\tfrac{1}{2}0^4 + 4^4 + \tfrac{1}{2}8^4] = 9216$$

$$= \tfrac{1}{2}A_8 + 4[4^4]$$

$$A_2 = 2[\tfrac{1}{2}0^4 + 2^4 + 4^4 + 6^4 + \tfrac{1}{2}8^4] = 7232$$

$$= \tfrac{1}{2}A_4 + 2[2^4 + 6^4] \quad \text{(less work)}$$

$$A_1 = 6724$$

$$A_{1/2} = 6596.25.$$

---

† Romberg Integration is sometimes called Richardson Extrapolation.

Now we can tabulate a sequence of even better approximations; explanations later.

$$A_8 = 16384$$

$$A_4 = 9216 \qquad A_4^* = \frac{4A_4 - A_8}{3} = 6826.67$$

$$A_2 = 7232 \qquad A_2^* = \frac{4A_2 - A_4}{3} = 6570.67 \qquad A_2^{**} = 6553.6$$

$$A_1 = 6724 \qquad A_1^* = \frac{4A_1 - A_2}{3} = 6554.67 \qquad A_1^{**} = 6553.6$$

$$A_{1/2} = 6596.25 \qquad A_{1/2}^* = \frac{4A_{1/2} - A_1}{3} = 6553.67 \qquad A_{1/2}^{**} = 6553.6$$

Thus, A_4^* assumes that the error is proportional to h^2, as shown in Fig. 6.3. Let's assume we can get the exact area by removing the error.

$$A \text{ ('exact')} = A_8 - k8^2 = A_4 - k4^2,$$

or, solving for k and substituting,

$$A \text{ ('exact')} = \frac{4A_4 - A_8}{3} = A_4^*$$

as in the table. Of course, this is less than fully successful; hence our quotes and the need for A_2^*, etc. Similarly, for the third column we assume that the remaining errors in the second column are proportional to h^4. If we were to

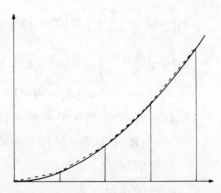

Fig. 6.3   In the Trapezoidal Rule, we model the area under a curve as the sum of the trapezia generated by the dotted line. The errors, and hence the error in area, are bad, and are proportional to h^2.

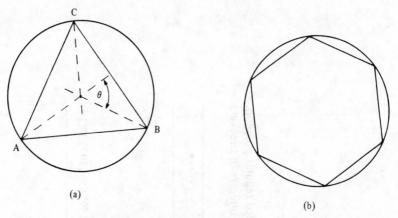

(a)

(b)

Fig. 6.4   In (a) triangle ABC is inscribed in a circle of unit radius. The triangle has area $3 \sin \theta \cos \theta$ where $\theta = \pi/3$. In (b) N is larger and the area tends to π.

introduce a fourth column, we should assume the error is proportional to h^6, and so on:

$$A_1^{**} = \frac{16 A_1^* - A_2^*}{15}$$

$$A_1^{***} = \frac{64 A_1^{**} - A_2^{**}}{63} \quad \text{etc.}$$

Let's now look at some amusing examples of this technique. Following Archimedes, we might estimate π by successively doubling the number of sides in a polygon inscribed in a unit circle, as in Fig. 6.4. (This involves solving a quadratic equation each time.) Examine Table 6.1 closely, and you will see for example that including $N = 12, 24, 48, 96$ would give only 10 decimal accuracy, whereas with $N = 3, 6, 12, 24, 48, 96$ we achieve 14 decimal accuracy. Thus the first two terrible estimates in fact contribute a great deal, in the end. Aren't numbers fascinating!?

Table 6.2 shows another case. Here, the first column has first-order convergence in $1/N$, the second has second-order convergence, and so on. [A generalised formula for A('exact') is

$$A(\text{'exact'}) = \frac{h_{i-1}^n A_i - h_i^n A_{i-1}}{h_{i-1}^n - h_i^n}$$

where n is the order of convergence and i refers to the value of N in the tables. Check our deductions for A_1^*, A_1^{**}, and A_1^{***}.]

The central question is, how do we know that we have enough values? The best answer in practice is trial and error. Even if we did not know the exact answers, it would be very obvious that Tables 6.1 and 6.2 are successful. The changes in the bottom row are very small at the end. So really, we guess, and often we guess correctly.

Table 6.1 Romberg integration to derive π as the limit of the area of an N-sided regular polygon. These numbers are extracted from a double-precision computer printout. The integers in brackets are the number of correct decimals.

N	A	A*	A**	A***	A****	A*****
3	1.299					
6	2.598	3.0311				
12	3.000	3.1340	3.1408			
24	3.106	3.1411(2)	3.1416(4)	3.14159191(5)		
48	3.133	3.1416(4)	3.1416(6)	3.14159265(8)	3.14159265347(9)	
96	3.139	3.1416(5)	3.1416(8)	3.14159265(10)	3.14159265359(13)	3.14159265358979 00(14)

Table 6.2 The first column is $A = (1 + 1/N)^N$ which tends towards e, with order N^{-1}. The second column is $A_i^* = 2A_i - A_{i-1}$ and the errors are of order N^{-2}. The third is $A_i^{**} = (4A_i^{**} - A_{i-1})/3$, etc. The convergence is less spectacular, because the orders of convergence are lower than in Fig. 6.1.

N					
1	2				
2	2.25	2.5			
4	2.44141	2.63282	2.67709		
8	2.56578	2.69015	2.70926	2.71386	
16	2.63793	2.71008	2.71672	2.71779	2.71805
32	2.67699	2.71605	2.71804	2.71823	2.71826	2.71827

6.7 EULER–MACLAURIN

Back to practical methods, to a trick which is very useful, very occasionally. If you look at Fig. 6.5 you will probably agree that the formula

$$\int_0^{Nh} f(x)\,dx = h[\tfrac{1}{2}f(0) + f(h) + f(2h) + \cdots + f(\overline{N-1}h) + \tfrac{1}{2}f(Nh)]$$

$$+ \frac{h^2}{12}[f'(0) - f'(Nh)]$$

$$- \frac{h^4}{720}[f'''(0) - f'''(Nh)]$$

$$\cdots \tag{6.7}$$

makes a sort of sense. As with Romberg, we start with a trapezoidal version, but then we correct it using the first, third, fifth ... derivatives at the two ends. Note we use derivatives at the ends only.

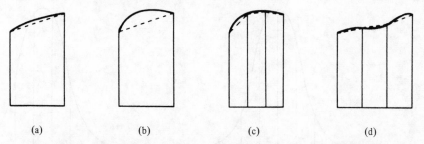

(a) (b) (c) (d)

Fig. 6.5 The error in the trapezoidal rule depends predominantly on the difference in slopes between the two extreme ends. In (a) the difference in slope is small whereas in (b) the difference and the error are larger. You can derive the second term of equation (6.7) by using a parabola: try it. In (c) the intermediate slopes cancel when we add the contributions: we are left with just the end slopes. In (d) the error in the middle section is of opposite sign, but less than the others combined. The formula accounts for this sort of behaviour.

The formula can be derived by the calculus of finite differences:

$$u_{n+1} = e^{hd/dx}u_n = Eu_n = (1 + \Delta)u_n$$

using the operator form of Taylor's series. (Section 2.2). Thus

$$\Delta = e^{hd/dx} - 1$$

$$\Sigma = \Delta^{-1} = \frac{1}{e^{hd/dx} - 1}.$$

We expand this, as a series! Fascinating, but it isn't easy to interpret what emerges. The integral comes from $(d/dx)^{-1}$. This derivation is not rigorous.

An alternative version of Euler–Maclaurin, based on the mid-ordinate rule, is even more attractive:

$$\int_0^{Nh} f(x)\,dx = h\{f(\tfrac{1}{2}h) + f(1\tfrac{1}{2}h) + \cdots f[(N - \tfrac{1}{2})h]\}$$

$$- \frac{h}{24}\{f'(0) - f'(Nh)\} + \frac{7h^3}{5760}\{f'''(0) - f'''(Nh)\}.$$

These coefficients can be found by expanding δ^{-1} from (3.3), section 3.4.

Euler–Maclaurin has been recommended in the past, if we want a table of integrals, of a function that we can differentiate but not integrate. Or, more frequently, if we have an infinite integral, and all the derivatives vanish at $\pm\infty$—the results are not exact, but are remarkably accurate: using equation (6.7) with $h = 1$, and noting the function is symmetrical about $x = 0$:

$$\int_{-\infty}^{\infty} e^{-(1/2)x^2}\,dx \cong 2[\tfrac{1}{2}1 + e^{-(1/2)} + e^{-2} + e^{-4(1/2)} + e^{-8}\cdots]$$

$$= 2.506628290$$

$$(\sqrt{2\pi} = 2.506628275 = \text{exact integral}).$$

Isn't this truly amazing? Numbers can be exciting.

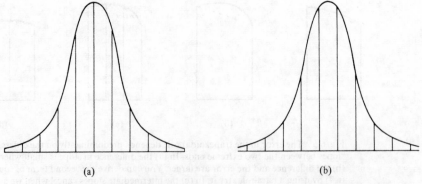

Fig. 6.6 The integral of a bell curve is very accurately represented by h ($\sum$ sample values) whatever the starting point. That is, the 'phase difference' between (a) and (b) normally makes little difference.

Euler–Maclaurin has gone out of fashion in the last generation. It used to be taught in statistics. We suspect that it has been dropped because pedantic teachers like to talk about 'orders of convergence' as in Romberg. But Euler–Maclaurin does much better. It seems to give hundreds of decimals of accuracy with $h = 0.2$, in the last case above.

Finally, then, Euler–Maclaurin is an obvious response to a task as in Fig. 6.6. 'Bell curves' are quite common in an engineer's life. Although the accuracy is less startling here than in the case of infinite integrals, this is a labour-saving trick that every practical engineer should know about.

6.8 MULTIPLE INTEGRATION

When trying to estimate the area of a park, the volume of water in a pond, or the quantity of gravel in a quarry we might end up trying to integrate with respect to more than one variable. How do we do this? Simple really: an example will show how it can be done. We will determine numerically the volume created when an ellipse of semi-axes a and b (in the x and y directions respectively) is rotated about the x-axis.

(We will not, at this stage, assign any numerical value to a and b).

We know the equation for such an ellipse is

$$\frac{x^2}{a^2} + \frac{y^2}{b^2} = 1$$

Thus, for a given value of x, $y = b\sqrt{1 - \left(\dfrac{x}{a}\right)^2}$.

Also, when an element of area $dx\,dy$ is rotated about the x-axis, it will rotate on radius y and thus create a volume $2\pi y\,dx\,dy$.

Hence, the integral we want is (see Fig. 6.7)

$$\text{Vol} = 2 \int_0^a \int_0^{b\sqrt{1 - (x/a)^2}} 2\pi y \, dy \, dx.$$

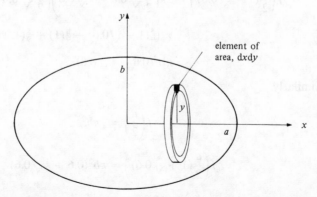

Fig. 6.7   An ellipse of semi-axes a and b is rotated about the x-axis. An element of area $dx\,dy$ creates a volume $2\pi y\,dx\,dy$ (and also covers the mirror area in the bottom of the ellipse).

We would obtain the exact solution by integrating first the inner integral

$$\int_0^{b\sqrt{1-(x/a)^2}} 2\pi y \, dy = \pi y^2 \Big|_0^{b\sqrt{1-(x/a)^2}} = \pi b^2 \left[1 - \left(\frac{x}{a}\right)^2 \right]$$

and then the outer integral

$$\text{Vol} = 2 \int_0^a \pi b^2 \left[1 - \left(\frac{x}{a}\right)^2 \right] dx = 2\pi b^2 \left(x - \frac{x^3}{3a^2} \right) \Big|_0^a = \frac{4}{3} \pi a b^2.$$

(Note as an aside and a check, that if $a = b$, we would have had a circle and integrated to find the volume of a sphere—correctly. *Always* use checks when you can.)

Now, use the Gauss 3-point rule to calculate the volume numerically. As before, we integrate the inside integral first, then the outside integral. If the inner integral gives us $f(x)$, we will need the three following values to do the outer integral.

$$f\left(\frac{a}{2}(1 - \sqrt{0.6})\right), \ f\left(\frac{a}{2}\right), f\left[\frac{a}{2}(1 + \sqrt{0.6})\right].$$

$$\left(\text{where } f(x) = \int_0^{b\sqrt{1-(x/a)^2}} \cdot 2\pi y \, dy\right).$$

So we calculate

$$f\left(\frac{a}{2}(1 - \sqrt{0.6})\right) = \int_0^{b\sqrt{1 - (1/4)(1 - \sqrt{0.6})^2}} 2\pi y \, dy = \int_0^{b\sqrt{0.6 - (1/2)\sqrt{0.6}}} 2\pi y \, dy$$

which from equation (6.4),

$$= \int_{-1}^1 f\left(\frac{b}{2}\sqrt{0.6 - \tfrac{1}{2}\sqrt{0.6}}(1 + \xi)\right) \frac{b}{2}\sqrt{0.6 - \tfrac{1}{2}\sqrt{0.6}} \, d\xi$$

where $f(y) = 2\pi y$

$$f\left(\frac{a}{2}(1 - \sqrt{0.6})\right) = 2\pi\left(\frac{b}{2}\sqrt{0.6 - \tfrac{1}{2}\sqrt{0.6}}\right)\left(\frac{b}{2}\sqrt{0.6 - \tfrac{1}{2}\sqrt{0.6}}\right)$$

$$\times \left(\tfrac{5}{9}(1 - \sqrt{0.6}) + \tfrac{8}{9}(1) + \tfrac{5}{9}(1 + \sqrt{0.6})\right)$$

$$= \pi b^2 (0.6 - \tfrac{1}{2}\sqrt{0.6}).$$

Similarly,

$$f\left(\frac{a}{2}\right) = \tfrac{3}{4}\pi b^2$$

$$f\left(\frac{a}{2}(1 + \sqrt{0.6})\right) = \pi b^2(0.6 + \tfrac{1}{2}\sqrt{0.6})$$

$$\therefore \quad \text{Volume} = 2\left(\frac{a}{2}\right)(\pi b^2)(\tfrac{5}{9}(0.6 - \tfrac{1}{2}\sqrt{0.6}) + \tfrac{8}{9}\tfrac{3}{4} + \tfrac{5}{9}(0.6 + \tfrac{1}{2}\sqrt{0.6}))$$

$$= \tfrac{4}{3}\pi a b^2.$$

Table 6.3 Abscissas and weight factors for Gaussian integration

$$\int_{-1}^{+1} f(x)\,dx = \sum_{i=1}^{n} w_i f(x_i)$$

Abscissas $= \pm x_i$ (Zeros of Legendre Polynomials)

Weight Factors $= w_i$

$\pm x_i$	w_i
n = 2	
0.57735 02691 89626	1.00000 00000 00000
n = 3	
0.00000 00000 00000	0.88888 88888 88889
0.77459 66692 41483	0.55555 55555 55556
n = 4	
0.33998 10435 84856	0.65214 51548 62546
0.86113 63115 94053	0.34785 84851 37454
n = 5	
0.00000 00000 00000	0.56888 88888 88889
0.53846 93101 05683	0.47862 86704 99366
0.90617 98459 38664	0.23692 68850 56189
n = 6	
0.23861 91860 83197	0.46791 39345 72691
0.66120 93864 66265	0.36076 15730 48139
0.93246 95142 03152	0.17132 44923 79170
n = 7	
0.00000 00000 00000	0.41795 91836 73469
0.40584 51513 77397	0.38183 00505 05119
0.74153 11855 99394	0.27970 53914 89277
0.94910 79123 42759	0.12948 49661 68870
n = 8	
0.18343 46424 95650	0.36268 37833 78362
0.52553 24099 16329	0.31370 66458 77887
0.79666 64774 13627	0.22238 10344 53374
0.96028 98564 97536	0.10122 85362 90376
n = 9	
0.00000 00000 00000	0.33023 93550 01260
0.32425 34234 03809	0.31234 70770 40003
0.61337 14327 00590	0.26061 06964 02935
0.83603 11073 26636	0.18064 81606 94857
0.96816 02395 07626	0.08127 43883 61574
n = 10	
0.14887 43389 81631	0.29552 42247 14753
0.43339 53941 29247	0.26926 67193 09996
0.67940 95682 99024	0.21908 63625 15982
0.86506 33666 88985	0.14945 13491 50581
0.97390 65285 17172	0.06667 13443 08688

Exact! Surprised? If you are, re-read sections 6.4 and 6.5 and consider the starting equation here.

There are other ways, of course. We could have considered the function as $f(x, y)$. The values of f at the specific Gauss points (e.g.: $(-\sqrt{0.6}, -\sqrt{0.6})$, $(-\sqrt{0.6}, 0)$ etc.) are the values required. Then we would add them all up with appropriate weighting factors.

EXERCISES

6.1. Derive the Gauss 2-point rule, and try it on a simple example.

6.2. Use the Gauss 3-point, Simpson, and the $7 - 32 - 12 - 32 - 7$ rule to estimate:

(i) π by $4 \displaystyle\int_0^1 \frac{dx}{1 + x^2}$

(ii) $\ln 2$ by $\displaystyle\int_1^2 \frac{dx}{x}$

6.3. Use the first five lines of the Romberg Integration technique to estimate

(a) $\displaystyle\int_0^\pi \sin^2 \theta \; d\theta$

(b) $\displaystyle\int_0^{\pi/4} \tan x \; dx$

and compare with the exact values.

6.4. Use Euler–Maclaurin to estimate

$$\int_{-\infty}^{\infty} e^{-x^4} \, dx$$

Try any other method.

6.5. Determine an expression for the integral

$$\int_{-1}^{1} \int_{-1}^{1} \phi(\xi, \eta) \; d\xi \; d\eta$$

using Gauss 3-point quadrature. Calculate a value for the integral if

$$\phi(\xi, \eta) = (1 - \xi^2)(1 - \eta^2)(1 - \xi\eta).$$

6.6. The shape of a cable under its own weight is given by

$$z = \frac{4wx}{l^2} (l - x) + \frac{hx}{l}$$

where the parameters are as shown.

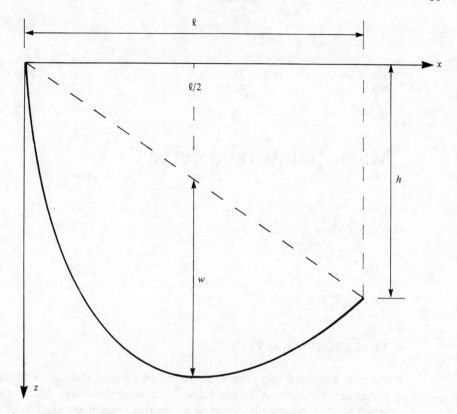

The length L of the cable is given by

$$L = \int_0^l \sqrt{1 + (z')^2} \, dx.$$

Using 'trial and error' or some other procedure, in conjunction with 3-point Gauss integration, determine w if

$$l = 60 \text{ m}, \ h = 40 \text{ m}, \text{ and } L = 100 \text{ m}$$

6.7. In an electrical circuit containing a capacitor C and a resistor R, the impressed voltage V is related to the current I by

$$V = RI + \frac{1}{C} \int_0^t I \, dt$$

Suppose $I = \ln(2t^2 + 1)$ amps in a circuit where $R = 0.1$ ohm and $C = 1$ farad. Find the voltage at 2 s.
  (i) Using the trapezoidal rule with $h = 0.2$
  (ii) Using the Simpson rule with $h = 0.2$
  (iii) Using the $7 - 32 - 12 - 32 - 7$ rule with $h = 0.25$
  (iv) Exactly.

7

Magic points on a curve

7.1 ERRORS AGAIN

Without enquiring too deeply, we ought to know a little more about the
properties of Lagrange interpolation. The magnitude and distribution of
errors is what principally concerns us. In this chapter we shall meet some
useful ideas, albeit without any strong mathematical foundation.

In section 2.3 we postulated that (a) a function can be expressed as a series
to x^{N-1} with tolerable error; and that (b) the series to x^N has virtually no
error. This frequently happens, and it forms the conceptual basis of the
present chapter.

We merge this idea with the experience of section 4.6: Lagrange interpolation seems to behave badly near the endpoints x_1 and x_N—a sort of
'whiplash' effect. Lagrange's interpolated function is exactly an $(N-1)$th
degree polynomial in x. The original function is, we postulate, nearly an
Nth degree polynomial. The difference—the error—is therefore another Nth
degree polynomial. We make the error zero at $x_1 \ldots x_N$; the function is
exact at the interpolating points, nowhere else. Thus we can write a formula
for the error:

$$\text{error} \cong \text{constant times } (x - x_1)(x - x_2) \cdots (x - x_N)$$

to a high degree of accuracy. This sort of curve is shown in Fig. 7.1(a).

With these background ideas, we set ourselves the problem of choosing
$x_1 \cdots x_N$ optimally, so as to get good interpolation accuracy. That is to say,
we propose to 'sample' the function $y(x)$ at $x_1 \cdots x_N$, in such a way as to
optimise the efforts of Lagrange. The quest is reminiscent of Gauss integration—see Fig. 7.1(b).

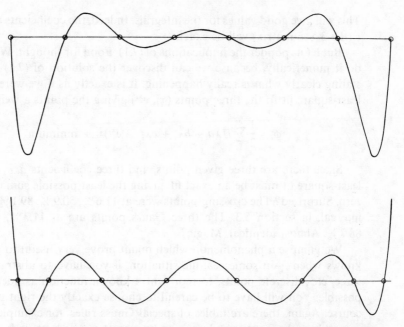

Fig. 7.1 (a) The expected form of the error, when interpolating a 9th degree polynomial through 8 equidistant points. With Lagrange, the errors don't always take the form expected, however; see Fig. 4.4 where sine is not a 9th degree polynomial. (b) Here we attempt to restrain the 'wildness' near x_1 and x_N by placing the points closer together near the ends, thus distributing the errors more evenly along the region of interpolation. Actually, we have used Gauss points.

7.2 THE GAUSS POINTS AS REPRESENTATIVE POINTS

Okay, the Gauss rules were a real success story (Chapter 6). And the success story extends to another phenomenon that every engineer should be aware of, far beyond the immediate task of integration. Suppose your problem is that you have to measure something, say a temperature, or a voltage, at a certain number of points within a region; what are the best points to choose? Of course, it depends on what your purpose is. But if the problem arises, you should start thinking creatively about Gauss points. More later.

Provocative example coming up. Reconsider the problem of section 5.3. Writing the Gauss rule as

$$\int_A^B f(x)\,dx = \sum_1^n H_i f(x_i)$$

we can do the necessary integrals for least-square smoothing of a function numerically, with

$$\sum_{i=1}^3 H_i \begin{bmatrix} 1 & x_i & x_i^2 \\ x_i & x_i^2 & x_i^3 \\ x_i^2 & x_i^3 & x_i^4 \end{bmatrix} \begin{Bmatrix} a \\ b \\ c \end{Bmatrix} = \sum_{i=1}^3 H_i \begin{Bmatrix} 1 \\ x_i \\ x_i^2 \end{Bmatrix} e^{x_i}. \qquad (7.1)$$

This will give good values for the integrals. Indeed, the coefficients of a, b, and c (the 3×3 matrix) will be exact.

But let us ponder the implications of (7.1). Food for thought. We need not do it numerically, because we can discover the solution of (7.1) merely by stating clearly what is really happening. It is exactly as if we were seeking a least-square fit to the three points (x_i, e^{x_i}) giving the points a weight H_i:

$$\phi = \frac{1}{2} \sum_1^3 H_i(a + bx_i + cx_i^2 - e^{x_i})^2 \rightarrow \text{minimum.}$$

Since there are three given points, and three coefficients a, b, and c, the least-square fit must be an exact fit, giving the least possible sum of squares, zero. Surprised? The crossing-points were at 11.6%, 50.9%, 89.1% along the interval, in section 5.3. The three Gauss points are at 11.3%, 50%, and 88.7%. Almost identical. Magic?

We glimpse a phenomenon which might prove very useful to you—who knows when?—in some real-life situation, if you have to abstract two, or three, etc., discrete points, to represent a known function as adequately as possible. You will have to be careful to choose exactly the right number, of course. Again, there are tables of special Gauss rules, for example to use in cases where the function is known to be zero at one or both ends, etc. When you get it right, the improvement in efficiency may be spectacular. You might be measuring a temperature, or a voltage, at some chosen points in a given domain. For example, within a finite element the stress varies appreciably. The best points to output stresses at are 'Barlow points', which are particular Gauss integration points in practice.

There is another way to justify what we are suggesting. Often we are not interested in the pointwise behaviour of a function. Its properties crystallise into a few numbers, which we call 'moments' $\int x^N y(x)\, dx$ over the range, where $N = 0, 1, 2$ and perhaps more. If from our sampled values we define the function as the interpolated polynomial, then the moments are exactly what we would calculate from the Gauss rule. That is, they should be good values.

7.3 CHEBYCHEV'S OSCILLATING POLYNOMIALS

Optimum points again: until now we have defined the aim in terms of minimising the mean square of the error. As we have seen, this still permits quite bad errors at the extreme ends of the range—although it is much better than with equally spaced x_i. Now we shall learn another technique, based on a different set of x_i, which minimise the greatest errors. To do this, we must learn first about a new set of functions.

Try to imagine the spot of light on a cathode ray tube. Suppose this is set up to respond to two alternating voltages in the x and y directions, $E_x = \cos \omega_x t$ and $E_y = \cos(\omega_y + \phi)t$. If ω_y is an exact multiple of ω_x, the curve that the point describes will repeat its path, and if the phase angle ϕ is zero, it will be a single curve, without loops. (For example, if $\phi = \pi/2$ and $\omega_x = \omega_y$ the curve will be a circle.) See Fig. 7.2.

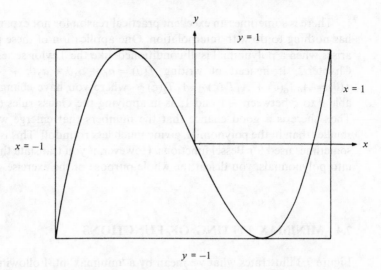

Fig. 7.2 The Chebychev cubic polynomial $y = T_3(x)$, is viewed here as $x = \cos \omega t$, $y = \cos 3\omega t$. It is a particular case of Lissajou's figures, which you may have seen in electronics or acoustics, where a point is oscillating simultaneously in the x- and y-directions.

Following Fig. 7.2 and putting

$$x = \cos \theta$$
$$y = \cos n\theta$$

then

$$y = T_n(x)$$

the nth degree 'Chebychev polynomial', if we can eliminate θ. We cannot do this directly, but it is easy to find a useful recurrence relation:

$$
\begin{aligned}
T_{n-1}(x) + T_{n+1}(x) &= \cos(n-1)\theta + \cos(n+1)\theta \\
&= (\cos n\theta \cos \theta + \sin n\theta \sin \theta) \\
&\quad + (\cos n\theta \cos \theta - \sin n\theta \sin \theta) \\
&= 2 \cos n\theta \cos \theta \\
&= 2 \times T_n(x).
\end{aligned}
$$

Thus $T_{n+1}(x) = 2 \times T_n(x) - T_{n-1}(x)$. Starting with $n = 0$, $\cos n\theta = y = 1$, so $T_0(x) = 1$ and $n = 1$, $\cos n\theta = \cos \theta = y = x$, so $T_1(x) = x$. Thus using our formula for $T_{n+1}(x)$ above, we have $T_2(x) = 2x^2 - 1$ (check with $n = 2$, $\cos n\theta = 2\cos^2\theta - 1 = 2x^2 - 1$: it agrees!)

$$T_3(x) = 4x^3 - 3x$$

etc.

However, don't ever solve this algebraically. The purpose of such a recurrence formula is numerical. For $x = 0.3$, we write $T_0(0.3) = 1$, $T_1(0.3) = 0.3$, hence we calculate $T_2(0.3) = 0.6 \times 0.3 - 1 = -0.82$, etc.

There is sometimes an excellent practical reason for not expanding, which has nothing to do with interpolation. One application of these polynomials arises when a polynomial is ill-conditioned, like the Taylor series for $\sin \theta$ in Chapter 2. If instead of writing $f(\theta) = a_0 + a_1\theta + a_2\theta^2 + \cdots$ we write $f(\theta) = A_0 T_0(\xi) + A_1 T_1(\xi) + A_2 T_2(\xi) \cdots$ where you have changed the variable θ to ξ between -1 and 1, as in applying the Gauss rules (section 6.4). Then there is a good chance that the numbers that emerge will be much smaller than in the polynomial, giving much less roundoff. This is a technique sometimes used for Bessel functions. However, if you translate the $T_n(\xi)$ back into polynomials, you defeat the whole purpose of the exercise.

7.4 MINIMAX FITTING OF FUNCTIONS

Figure 7.3 illustrates what we mean by a 'minimax' fit. Following the above paragraph, it may be possible to express $f(\theta)$ between $\theta = a$ and $\theta = b$ as $A_0 T_0(\xi) + A_1 T_1(\xi) + \cdots$ where $\theta = \frac{1}{2}(a + b) + \frac{1}{2}\xi(b - a)$, as in section 6.3. In general, the A_n will form a series converging to zero. If we are lucky, $A_0 T_0(\xi) + \cdots + A_n T_n(\xi)$ gives negligible errors. In that case, we can remove $A_n T_n(\xi)$ and generate a minimax fit with $(n + 1)$ equal errors having alternating sign. For the error will be $A_n T_n(\xi)$, which oscillates between A_n and $-A_n$ as ξ goes from -1 to $+1$.

Some numerical analysis courses invite you to take a polynomial, to express it as a linear combination of Chebychev polynomials, then to strike out the highest so as to generate an 'exact' minimax fit. (There is a simple technique for finding the A_i directly from the function, by numerical integration.) If the A_i converge well, it should be possible to find a minimax approximation but not of any required degree. This is a peculiarly useless exercise. In our opinion, such a task will in practice involve trial and error. As

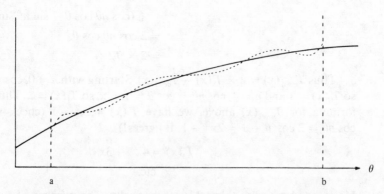

Fig. 7.3 Diagrammatically, the dotted curve represents the approximation to the given function. In a 'minimax' fit, the errors are alternately positive and negative, and equal in magnitude.

a first trial, the best approach will be to put the approximating function through the N zeros of $T_N(\xi)$, i.e.

$$\cos n\theta = 0, \ \theta = (n + \tfrac{1}{2})\frac{\pi}{N}, \ n = 0 \cdots (N-1)$$

$$\xi_n = \cos(n + \tfrac{1}{2})\frac{\pi}{N} \tag{7.2}$$

$$x_n = \tfrac{1}{2}(a + b) + \tfrac{1}{2}(b - a)\cos(n + \tfrac{1}{2})\frac{\pi}{N}.$$

The errors will of course not be equal—we are never that lucky. Using trial and error, you could replace ξ_n by $\xi_n + K_1(1 - \xi_n^2) + K_2\xi_n(1 - \xi_n^2)$ for example, where K_1 and K_2 are constants chosen to give more nearly equal errors.

Thus, the Gauss sampling points would give Lagrange interpolation nearly equivalent to the much more laborious least-squares smoothing. The Chebychev zeros should give the approximate minimax fit. Note that the minimax fit really has much less justification than the least-squares fit.

7.5 MINIMAX AND LEAST-SQUARES

Embarrassing. We have not one optimum fit, but two! Not one set of magic points, but two! Can we compare them? Yes; in a general, intuitive way, and also in a slightly more precise, descriptive way. In Fig. 7.1(b) we saw that least-squares (Gauss points) gave the greatest errors at the ends. Those in the middle of the range were rather smaller. The difference was reduced, it was much less than with the equally spaced points, in Fig. 7.1(a). That is why these particular magic points tend to cluster near the ends. When we came to the minimax magic points, the Chebychev zeros, we insisted that even at the ends the errors were no more than those in the middle. So you would expect

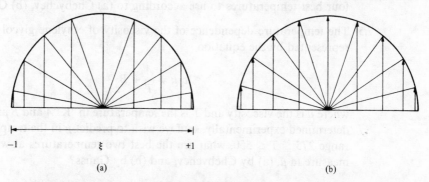

(a)   (b)

Fig. 7.4 (a) Construction of ξ, for Chebychev points from equation (7.2) with $N = 9$. Note, in this type of construction, the intermediate angles are all the same with the end angles precisely half in size. (b) The same construction for the Gauss points. The end angles are rather more than half of the others, which are as nearly equal as you could tell from a drawing.

the Chebychev points to cluster even more closely towards the ends, towards $\xi = \pm 1$.

Figure 7.4 illustrates these vague and fairly obvious statements. What is genuinely surprising is that the angles corresponding to the Gauss points are nearly equal—for there is nothing in the derivation that leads us to expect this. But what is really startling about Fig. 7.4(b) is that the vertical distances, the heights of the intersections above the base line, are almost exactly proportional to H_i, the Gaussian weighting factors. Thus with the semicircles of Fig. 7.4 we once again come full circle.

EXERCISES

7.1. The Chebychev polynomial recurrence relationship is

$$T_{n+1}(x) = 2xT_n(x) - T_{n-1}(x).$$

Deduce the roots of $T_6(x)$ and calculate the value of $T_6(0.5)$. Why does roundoff damage not accumulate catastrophically in any practical case? Hint: if r (section 3.2) is complex, $|r| \leq 1$ for stability.

7.2. Find an approximate cubic minimax fit to $\cos \theta$ (θ in radians), $-1 < \theta < 1$. (By symmetry, $A + B\theta^2$). Find the greatest errors.

7.3. Experiment with our program LAGRAN to see how good the Gauss points are as sampling points. For 8 points in the region $-1 < \xi < 1$ the points are at ± 0.18343464, ± 0.52553241, ± 0.79666648, ± 0.96028986. Try also the Chebychev zeros, calculated from equation (7.2).

7.4. The heat capacity of ammonia can be correlated with temperature through a polynomial expression. The expression requires four parameters which can be 'calibrated' by experiment. If we wish to calibrate the equation for the temperature range $200 \leq T \leq 1400°K$, what are the four best temperatures to use according to (a) Chebychev, (b) Gauss?

7.5. The temperature dependence of the viscosity of ethylene glycol can be represented by the equation

$$\ln \mu = \frac{A}{T} + B$$

where μ is the viscosity and T is the temperature in $°K$. A and B are to be determined experimentally, so if we wish to predict μ in the temperature range $275 \leq T \leq 550$, what are the best two temperatures at which to measure $\ln \mu$, (a) by Chebychev, and (b) by Gauss?

8

Differential equations

8.1 THE WAY TO THE STARS?

You have probably done a course in mathematics in which you solved various differential equations analytically. In the real world such tasks are usually done numerically, as with integration. Of course, this means that computers are used quite frequently for solving differential equations, especially in some engineering offices.

We timed it all rather nicely, we humans. Early science fiction writers like H. G. Wells really got it all wrong, this space travel. If their engineers had had access to modern rockets, with the necessary power, they would have succeeded—in a sense: they would have created dozens of useless moon satellites—for even now, who wants to communicate via satellite from one point on the moon to another point on the moon?

The difficulty of hitting the moon, let alone of landing on some precise spot, was simulated at the time of the first moon flight by a raised track of curious shape, so that if you threw a ball onto it, the ball described the path that a free rocket would take. It was impossible to hit the moon! Probably there were always a few specialists who appreciated the extreme difficulty of navigating a spacecraft as the flight proceeded—who appreciated that advanced rockets were not enough, and that computers would have to be invented before a journey to the moon could actually succeed! We timed it all rather nicely.

Yet a computer is not very smart, in itself. Only quick. In anticipating the flight of a rocket, it proceeds in steps of a minute or so, and calculates what happens at each time step, from what happened in previous time steps. It calculates the flight path, probably to about 15 decimals! So it has to know, at the present time step, the precise velocity of the rocket, its position, its present mass (some of the fuel has been burnt), the gravitational field, the rocket thrust (if any), and the wind resistance (if any). To plot our future path of study let's consider a simple example: one we can solve explicitly (so we can check our numbers!).

Problem

$$\frac{dx}{dt} = \text{function; find } x.$$

Solution:
Create a table, $x(t_o)$, $x(t_0 + \varepsilon)$, $x(t_0 + 2\varepsilon)$, ..., by calculating the value at each step from the previous step; integrating both sides, and introducing limits,

$$\int_t^{t+\varepsilon} (\text{function}) \, dt = x \Big|_t^{t+\varepsilon} = x(t + \varepsilon) - x(t).$$

Thus

$$x(t + \varepsilon) = x(t) + \int_t^{t+\varepsilon} (\text{function}) \, dt$$

$$= x(t) + \varepsilon\{\text{mean value of function}\}.$$

Precisely the same problem as numerical integration, Chapter 6!

Snag:
Consider

$$\frac{dx}{dt} = x + t, \, x(1) = 1$$

(Exact solution $x = 3e^{t-1} - t - 1$).

In going from $t = 1$ to $t = 1 + \varepsilon$ you can't calculate the function, because you don't yet know x at the sampling points in the interval t to $t + \varepsilon$. So you can't possibly find the mean value. Shucks.

Solution:
Regard the techniques that you will learn as 'experiments'. The computer makes exploratory forays into the unknown territory, between t and $t + \varepsilon$. With the intelligence it brings back, it then estimates—very accurately—the mean velocity dx/dt, so as to satisfy the equation best.

8.2 A CRUDE NUMBER-CRUNCHING SOLUTION

In industry, we often use the enormous arithmetical capability of computers to grind out some sort of answer with the minimum of thought. For example, the temperature θ along the rod in Fig. 8.1 is controlled by the equation

$$K \frac{\partial^2 \theta}{\partial x^2} = \mu \frac{\partial \theta}{\partial t} \tag{8.1}$$

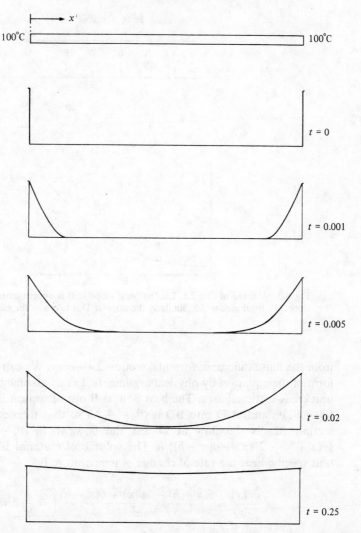

Fig. 8.1 Thermal conduction, along a uniform bar. At $t = 0$, the temperature is zero; but the ends are instantaneously raised to 100°C. Eventually the temperature becomes 100°C throughout, because no heat is lost. An exact solution by Fourier series is known.

where K is the conductivity and μ is the specific heat. For simplicity, we make both unity. Because this is a 'partial' differential equation (∂ instead of d—see section 2.2) we must take steps along the rod in space, as well as ε in time.

In the solution to (8.1) we shall need an extensive table, for $x = 0, 0.1, 0.2$ ... 1.0 say, across the page, giving a row for each time step, $t \cong 0, 0.001, 0.002$, etc., indefinitely, until we run out of paper. If $h = 0.1$, the step in x, we can write the left-hand side of (8.1) for any row as

$$\frac{\partial^2 \theta}{\partial x^2} = \frac{1}{h^2} \delta^2 \theta = \frac{1}{h^2} \{\theta(x+h) - 2\theta(x) + \theta(x-h)\}$$

Differential Equations

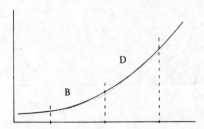

Fig. 8.2   A detail of Fig. 8.1. The 'halfway' slope at B is approximately the 'mean' slope AC, from section 2.3. Similarly the slope at D is taken as the mean slope CE.

from the finite difference formula, section 2.4—easy. We can also derive this form of the equation by physical arguments. Let us assume that the bar has unit cross-sectional area. The heat flow at B out of section BD in Fig. 8.2 is $(\theta_C - \theta_A)/h$ and at D into BD is $(\theta_E - \theta_C)/h$ so that the net heat flow into section BD is the flow at D less the flow at B, $(\theta_E - 2\theta_C + \theta_A)/h =$ $\{\theta(x + h) - 2\theta(x) + \theta(x - h)\}/h$. The volume of material BD is h, so with unit specific heat the rate of change of temperature is

$$\frac{\partial\theta(x)}{\partial t} = \frac{\theta(x + h) - 2\theta(x) + \theta(x - h)}{h^2} \left(= \frac{1}{h^2}\delta^2\theta \right).$$

Our assumption in the t-direction is very crude, by comparison. We calculate $\partial\theta/\partial t$ at each x, and then we take this as the mean value, between t and $t + \varepsilon$; i.e. we assume it doesn't change at all in the time interval; it retains its starting value. Our excuse is that we have to take very short time steps, for reasons that will appear later.

Table 8.1 needs a little explanation. For example, in the eighth line, $55 = \delta^2\theta = 100 - 2 \times 24 + 3$, from the finite difference formula, section 2.4, and in the ninth line, $30 = 24 + (55/h^2)\varepsilon$: $(55/h^2)$ is $(\delta^2\theta/h^2)$ which we have shown to be $\partial\theta/\partial t$ for this x at this t; $\varepsilon = 0.001$, $h = 0.1$. We have programmed this method (formally, Euler's method) in THERM, where you can compare with the exact solution formulated in our subroutine CHECK. The latter is based on the Fourier analysis of the starting condition. Then each term $\cos \alpha x$ gives rise to the transient $\exp(-K\alpha^2 t/\mu)\cos \alpha x$. There is no obvious virtue in your re-programming this.

Table 8.1 Temperature of positions in the rod of Fig. 8.1 with increasing time, using a small time step. The result is stable and acceptably accurate.

t	x	0	0.1	0.2	0.3	0.4	0.5	0.6	0.7	0.8	0.9	1.0
0	θ	100	0	0	0	0	0	0	0	0	0	100
	$\delta^2\theta$	—	100	0	0	0	0	0	0	0	100	—
0.001	θ	100	10	0	0	0	0	0	0	0	10	100
	$\delta^2\theta$	—	80	10	0	0	0	0	0	10	80	—
0.002	θ	100	18	1	0	0	0	0	0	1	18	100
	$\delta^2\theta$	—	65	16	1	0	0	0	1	16	65	—
0.003	θ	100	24	3	1	0	0	0	1	3	24	100
	$\delta^2\theta$	—	55	18	3	0	0	0	3	18	55	—
0.004	θ	100	30	5	3	0	0	0	3	5	30	100
	$\delta^2\theta$	—	45	20	5	0	0	0	5	20	45	—
0.005	θ	100	34	7	1	0	0	0	1	7	34	100

Table 8.2 Temperature of positions in the rod of Fig. 8.1 with increasing time, using too large a time step: poor results.

t	x	0	0.1	0.2	0.3	0.4	0.5	0.6	0.7	0.8	0.9	1.0
0	θ	100	0	0	0	0	0	0	0	0	0	100
	$\delta^2\theta$	—	100	0	0	0	0	0	0	0	100	—
0.01	θ	100	100	0	0	0	0	0	0	0	100	100
	$\delta^2\theta$	—	−100	100	0	0	0	0	0	100	−100	—
0.02	θ	100	0	100	0	0	0	0	0	100	0	100
	$\delta^2\theta$	—	200	−200	100	0	0	0	100	−200	200	—
0.03	θ	100	200	−100	100	0	0	0	100	−100	200	100
	$\delta^2\theta$	—	−400	500	−300	100	0	100	−300	500	−400	—
0.04	θ	100	−200	400	−200	100	0	100	−200	400	−200	100

8.3 NUMERICAL INSTABILITY

Spot checks show that the results of this crude calculation are quite good. 'Instability', which we look at next, has nothing to do with the accuracy of the finite difference formulae. It has much in common with the runaway roundoff of section 3.2—when it happens, you simply have to start again with a shorter time step. Sad, however: it happens to the most enterprising and adventurous workers. We agree, the table of θ is progressing painfully slowly. But to show what would happen, let's speed it up by a factor of 10, Table 8.2:

Oh, dear. The zigzags in θ are beginning to expand about 2:1 at every step, and the computer will overflow very soon.

Zigzags. This observation tells us how to estimate roughly the maximum time step that we can use. What would happen to the following pattern in θ?

t	θ	...	1	-1	1	-1	1	...
	$\delta^2\theta$	...	-4	4	-4	4	-4	...
$t + \varepsilon$	θ	...	z	$-z$	z	$-z$	z	...

where $z = 1 - 4\varepsilon/h^2$. If $z < -1$, we have instability (the values of θ would be growing further apart rather than smoothing out), i.e. instability if $\varepsilon > h^2/2$. Hence, the maximum ε for stability.

Not only crude assumptions like ours give instability. The best formulae have this kind of limitation. We have emphasised it, because if you get involved with differential equations it is always a danger. It has nothing to do with accuracy. You took a bet and lost.

8.4 THERMAL CONDUCTION, UNIFORM PLATE

We can regard a plate as a collection of rectangular bars parallel to the x-axis, coexisting with a collection of bars parallel to the y-axis. That is, we have a buildup of heat flow in the x-direction, and we have at the same time a buildup of heat flow in the y-direction, both causing $\partial\theta/\partial t$. As a result, the total rate of temperature increase is the sum of the two contributions:

$$\frac{\partial\theta(x, y)}{\partial t} = \frac{\partial^2\theta}{\partial x^2} + \frac{\partial^2\theta}{\partial y^2} \tag{8.2}$$

$$= \frac{\theta(x + h, y) - 2\theta(x, y) + \theta(x - h, y)}{h^2}$$

$$+ \frac{\theta(x, y + h) - 2\theta(x, y) + \theta(x, y - h)}{h^2}.$$

(The right-hand side of (8.2) is very familiar in applied mathematics. It is often written as $\nabla^2\theta$. For example, the steady-state solution to the problem of thermal conduction occurs after a long time, when the transient has died away, and $\partial\theta/\partial t = 0$ everywhere: thus $\nabla^2\theta = 0$.)

The procedure for solving this is exactly the same as for the bar: except that it uses a lot more paper—an array in x and y for each time step.

Moreover, we need even shorter timesteps for a given h than in the case of the bar, section 8.2, to avoid instability. We argue thus: a pattern of $\theta(t)$

$$\cdots \quad \begin{matrix} 1 & -1 & 1 & -1 & 1 & -1 \\ -1 & 1 & -1 & 1 & -1 & 1 \\ 1 & -1 & 1 & -1 & 1 & -1 \\ -1 & 1 & -1 & 1 & -1 & 1 \end{matrix} \quad \cdots$$

would generate $\nabla^2\theta = \pm 8/h^2$ so that each $\theta = $ '1' becomes at $t + \varepsilon$ (the next time step) $\theta = 1 - 8\varepsilon/h^2$ which for stability must not be less than -1. Solving this gives $\varepsilon < \frac{1}{4}h^2$. The time step is small—very expensive. However, calculations as unappetising as this (and much more complicated) are very frequently done in industry. One learns to hold one's nose, when the answers are genuinely useful, and the boss seems happy.

8.5 HEUN

Here is a more accurate, more sophisticated method; as promised in our opening remarks, we 'explore' the next interval, before we jump. Thus, working two steps, with $\varepsilon = 0.03$, for the equation $dx/dt = x' = x + t$ with $x(1) = 1$:

In the layout of Table 8.3, the 'accurate' values are underlined; alternate columns are explorations, intermediate calculations, phoneys. And the underlined values are quite good—the exact solution, $3\exp(t - 1) - 1 - t = 1.125510$ for $t = 0.06$. We calculate each x' from the t and x above it, e.g. $2.06 = 1.02 + 1.04$. The exploratory columns (with no underlined values) penetrate $2/3\ \varepsilon$ into the next time step, e.g. from $t = 1.06$ the next trial would be at 1.08. We calculate each x from the previous underlined value, using a different slope for the exploratory and the underlined values. For the final 'weighted mean' slope, we take 25% of the initial slope, plus 75% of the 'phoney' slope. Consider now the second step, starting at $t = 1.03$.

(i) Calculate the starting slope, from the formula,

$$x + t = 1.06135 + 1.03 = 2.09135.$$

(ii) Use this slope to estimate the exploratory x at $t = 0.05$, (an increase in t of 0.02)

$$x = 1.06135 + 0.02 \times 2.09135 = 1.1031773.$$

Table 8.3 Position (x) versus time for $dx/dt = x + t$ using Heun's method.

t	1	1.02	1.03	1.05	1.06
x	1	1.04	1.06135	1.103177	1.125482
x'	2	1.06	2.09135	2.153177	2.185482

(iii) With this exploratory value, compute the 'phoney' slope,

$$x' = x + t = 1.103177 + 1.05 = 2.153177.$$

(iv) Compute the weighted mean slope, from these two slopes:

$$\tfrac{3}{4} \times (\text{'phoney' slope}) + \tfrac{1}{4} \times (\text{initial slope}) = 0.75 \times 2.153177$$

$$+ \, 0.25 \times 2.09135 = 2.137720.$$

(v) Compute the next underlined x, the previous underlined value plus (the weighted mean slope multiplied by the time interval).

$$1.06135 + 0.03 \times 2.137720 = 1.125482$$

In summary, the table is formed as shown in Table 8.4.

Table 8.4   Heun in symbolic notation.

t	t_0	$t_0 + \tfrac{2}{3}\varepsilon$	$t_0 + \varepsilon$
x	a	$c = a + \tfrac{2}{3}\varepsilon b$	$e = a + \varepsilon(\tfrac{1}{4}b + \tfrac{3}{4}d)$
x'	b	d	(You have to know how to calculate d (and b if not given) from t and x above: i.e. the relationship between x', x, and t.)

Memorise these rules. The tabular layout makes sense, and there is considerable educational value in working a few simple examples. This is quite interesting, and having done it by hand we are in a better position to program it. There is no virtue whatsoever in substituting, line after line, into the formulae from some book, without understanding the purpose of each step.

8.6   HEUN, SEVERAL VARIABLES

We can process two or more variables in parallel. For example, we could reduce the second-order equation $x'' - 2x' + x = 0$ to the two first-order equations.

$$\left. \begin{array}{l} x' = y \\ y' = 2y - x \end{array} \right\}. \tag{8.3}$$

The table is as before, except that we have twice as many entries†, Table 8.5. Check our maths.

---

† As far as we can discover, this idea was first suggested in 1950. There was a flurry of activity, as people began to see the possibilities of the computer.

Table 8.5   Layout for Heun for two
variables or a 2nd-order equation.

t	0	0.02	0.03
x	0	0.02	0.03090
y	1	1.04	1.06135
x'	1	1.04	
y'	2	2.06	

In truth, the exact solution $x = te^t$ gives 0.030914 and 1.061368. This is
disappointing. Don't despair, however. There are better methods than
Heun's.

8.7   RUNGE-KUTTA FOURTH-ORDER

This method is somewhat more complicated, but much more accurate.
Solving the same problem as in Table 8.3 again:

Table 8.6   The problem of Table 8.3 using Runge-Kutta.

t	1	1.1	1.1	1.2	1.2	1.3	1.3	1.4	1.4
x	1	1.2	1.23	1.466	1.4642	1.73062	1.767262	2.077652	2.075454
x'	2	2.3	2.33	2.666	2.6642	3.03062	3.067262	3.477652	

This example is contrived to give simple numbers in the first step. The mean
slope is $\{2 + 2(2.3 + 2.33) + 2.666\}/6 = 2.321$, hence the new underlined
value $1 + 0.2 \times 2.321 = 1.4642$. (Accurate value 1.464208). The accurate
value at $t = 1.4$ is 2.075454. The symbolic summary, like the one for Heun, is
in Table 8.7.

Table 8.7   Runge-Kutta shown symbolically.

t	t_0	$t_0 + \frac{1}{2}\varepsilon$	$t_0 + \frac{1}{2}\varepsilon$	$t_0 + \varepsilon$	$t_0 + \varepsilon$
x	a	$c = a + \frac{1}{2}\varepsilon b$	$e = a + \frac{1}{2}\varepsilon d$	$g = a + \varepsilon f$	$k = a + \varepsilon\underbrace{(b + 2d + 2f + h)/6}$
x'	b	d	f	h	mean slope

Thus each x' is based on the x and t above it, so that d for example would
involve c and $t_0 + \frac{1}{2}\varepsilon$. Each x is based on the latest underlined x, and the x' in
the preceding column. The final mean x' is reminiscent of Simpson's rule, with
the value for $t = \frac{1}{2}\varepsilon$ 'duplicated'. The advantage of this tabular layout is that
each column makes approximate sense, thus revealing any bad arithmetical

mistake. It is easy to remember, too†. We end with an example of a higher order differential equation

$$x''' = 3(x + t), \text{ with } x(1) = 1, x'(1) = 2, x''(1) = 3$$

which reduces to three simultaneous equations

$$x' = y$$

$$x'' = y' = z$$

$$x''' = z' = 3(x + t).$$

Thus

Table 8.8   Runge–Kutta for a fourth-order differential equation. The full table in (a), a telescoped version in (b) to save writing.

(a)

t	1	1.1	1.1	1.2	1.2	
x	1	1.2	1.23	1.472	1.4686	(exact answer 1.468626)
y	2	2.3	2.36	2.738		
z	3	3.6	3.69			
x'	2	2.3	2.36			
y'	3	3.6	3.69			
z'	6	6.9	6.99			

(b)

t	1	1.1	1.1	1.2	1.2
x	$\frac{1}{2}$	1.2	1.23	1.472	1.4686
$x' = y$	2	2.3	2.36	2.738	
$x'' = y' = z$	3	3.6	3.69		
$x''' = z'$	6	6.9	6.99		

We need not labour the point: Runge–Kutta is remarkably accurate.

8.8   YOU IN INDUSTRY

What should we teach you? Almost certainly, you will have an efficient 'black box' to solve your equations. There are many methods. A type we have not mentioned is the 'predictor-corrector'. It estimates roughly the next x_i using previous values in the table, then it improves the value and, incidentally, minimises the chance of instability. Such methods cost about half as much in

---

† Looking at contemporary textbooks, it would appear that during the 1950's this formula was selected from many others, because it is easy to remember. At that time an engineer (male) depended utterly on a computer (female) who pounded a mechanical calculator for the whole of her working day. Clarity of tabulation was essential. Now we have both male and female engineers, and computers are...

computer time, but they have to be started with a Runge–Kutta method. If you are caught without a black box, Runge–Kutta should suffice, and it is easy to program. If you need more accuracy, or if the computer goes berserk (unstable), start again with a shorter time step. Don't be unduly afraid of differential equations: ordinary equations can be equally spiteful .....

EXERCISES

8.1. A transient heat conduction problem is idealised below:

$$
\begin{array}{ccccc}
 & 10 & 10 & 10 & \\
9 & 4 & 3.5 & 4 & 9 \\
8 & 3 & 0 & 3 & 8 \\
7 & 2 & 1 & 2 & 7 \\
 & 6 & 6 & 6 &
\end{array}
$$

The mesh consists of unit squares. The values 6, 7, 8, 9, and 10 are kept constant. If

$$\frac{\partial^2 \theta}{\partial x^2} + \frac{\partial^2 \theta}{\partial y^2} = \frac{\partial \theta}{\partial t}.$$

Use Euler's method to estimate values of θ at the internal nodes at a time 0.01 later. At what time step, approximately, would the process become unstable?

Compare the results of 10 of the above steps with one of 0.1, 20 with one of 0.2 and 30 with one of 0.3.

8.2. Given
$x' = tx$, $x(1) = 1$; use your calculator to find $x(1.03)$ by Heun
$x' = x + t$, $x(0) = 1$; use your calculator to find $x(0.03)$ by Heun
$x' = t/x$, $x(1) = 0.5$; use your calculator to find $x(1.06)$ by Heun
—try both one and two steps for the latter case.

8.3. Given
$x' = tx$, $x(1) = 1$; use your calculator to find $x(1.2)$ by Runge–Kutta
$x' = x + t$, $x(0) = 1$; use your calculator to find $x(0.2)$ by Runge–Kutta
$x' = t/x$, $x(1) = 0.5$; use your calculator to find $x(1.1)$ by Runge–Kutta.

8.4. Use your calculator and Runge–Kutta to estimate $x(0.2)$ if
  (i) $x(0) = 0$, $x'(0) = 1$, $x''(0) = 2$, $x'''(0) = 1$, and $x'''' = d^4x/dt^4 = x$.
  (ii) $x(0) = 0$, $x'(0) = 1$, $x''(0) = 0$, $x'''(0) = 4$, and $x'''' = d^4x/dt^4 = 16x$
  (iii) $x(0) = 0.5$, $x'(0) = 2$, $x''(0) = -1$, $x'''(0) = 5$, and $x'''' = d^4x/dt^4 = 5d^2x/dt^2 - 4x$.

8.5. Use our programs HEUN and RUNGE to check your answers and do more steps. Determine the size of time step to give you satisfactory accuracy.

8.6. The equation of a pendulum with a large angle of swing (such as may be found in some gyroscopes) is:

$$\frac{d^2\theta}{dt^2} + \frac{mgl}{I}\sin\theta = 0$$

where t = time

θ = the angle from the vertical

m = mass (here 4.1207 kg)

l = length of suspension to centroid of mass (here 0.25 m)

I = moment of inertia about suspension point (here 0.261666 kg m^2)

g = gravitational acceleration (9.8067 m/s^2)

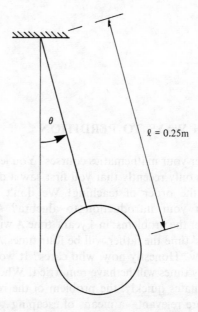

Use Runge–Kutta of order 4 to compute θ 2.5 seconds after release of the pendulum from an angle of 25°. Compare your answer with the value obtained using the approximation of $\sin\theta = \theta$ and 'exact' solution to the resulting differential equation.

9

Equations

9.1 TEN WAYS TO PERDITION

Remember your mathematics courses? You learnt to solve equations as kids, yet it was only recently that you first saw a differential equation? Yet we've reversed the order of teaching! We don't think we're mad.... Do you remember your introduction to algebra? At the time, it was puzzling. Remember the problems: in 4 years' time A will be a third as old as his father; in 2 years' time the father will be four times A's present age. How old is A's father now? Honestly now, who cares? It would be much more interesting, how many times will he have remarried! Whereas given access to something that calculates quickly, the problem of the rocket, in section 8.1, would be much more relevant—a means of escaping.... Perhaps this is the way that mathematics will be taught, one day. Tackle the easy things first. As far as possible, get involved with real problems.

Another approach. You have solved a differential equation, and you proudly present the 'solution': $\cos x + \tan x = \ln(t + t^2)$. Your math teacher says that's the genuine 'analytic solution'. The question does *not* continue, 'now plot x against t'. Thank goodness! Could you do it? The answers don't just appear! It is to filling such regrettable gaps in your education that this chapter is dedicated.

You have the impression that nothing could be easier! Besides, you have belatedly learnt to program your pocket calculator? Okay, program

$$f(\theta) = e^{-\tan^2\theta} = 0$$

and try various values of θ. Of course, $90°$ and $270°$ should give zero. But on our calculator, sometimes they do, sometimes they don't! And apart from $90°$,

102

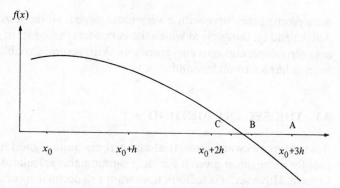

Fig. 9.1 Interval halving. A simple program to find roots. It takes uniform steps of size h, until $f(x)$ changes sign, first noted here at A. Then it comes back $\frac{1}{2}h$, to B, and discovers that $f(x)$ is still negative. So it comes back another $\frac{1}{4}h$ to C; $f(x)$ is now positive. It concludes that there is a root between B and C, so it goes *forward* by $h/8$. After a number of further interval halvings, the change in x becomes less than the roundoff tolerance on x; it has then found the root, for practical purposes. Mathematically, the 'Bisection Method'.

any angle between 86.3° and 93.7° gives zero. Any angle between 77.9° and 102.1° gives less than 5×10^{-10}, i.e. zero, to nine decimals!

Or perhaps you are the proud owner of an expensive calculator, with a special device to find roots? Some such calculators use progressive interval-halving to discover precisely (i.e. to the last place of decimals) where $f(x)$ changes sign—see Fig. 9.1. If the function is $\tan \theta$, it tells you that 0° and 180° are roots, okay, but it also tells you that 90° is a root! (Where plus infinity changes into minus infinity.) Sense of humour. And $\exp\{-\tan^2 \theta\}$ will generate no 'zeros' at all, even though we have seen it so generously endowed, because it never changes sign! Pocket calculators can be as stupid as computers.

9.2 ROUNDOFF

A glance at Fig. 9.2 shows what is happening with $\exp(-\tan^2 \theta)$. To a lesser degree, this can happen with more ordinary functions. Because of roundoff, your calculator describes an indeterminate 'path' of finite width, as if you

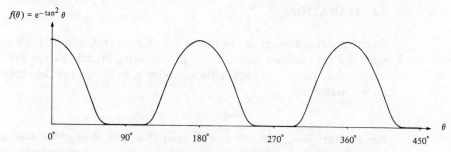

Fig. 9.2 Is this a cosine curve that has melted in the sun? Problem: $f(\theta) = 0$. Answer: $\theta = 78°$ to 102°??? (9 decimals).

were plotting the curve with a very blunt pencil. So the haziness of the root will depend on the angle at which the curve cuts the *x*-axis in general—in this case zero slope, and zero curvature too. With several variables, this phenomenon is harder to understand.

9.3 THE SECANT METHOD

Apart from awkward cases, trial-and-error is quite a good method to find a root for an engineer given a simple programmable calculator, and the ability to use it. However, it is tedious, if we want 6–8 decimal accuracy, and you will find the secant method much quicker. This assumes that the function is linear, very locally, as in Fig. 9.3, hence

$$x = \frac{af(b) - bf(a)}{f(b) - f(a)}. \tag{9.1}$$

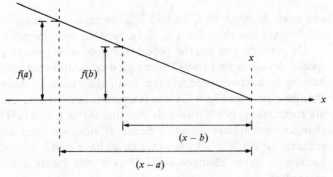

Fig. 9.3 Finding the roots of $f(x) = 0$, assumed linear. By similar triangles, $(x - a)/f(a) = (x - b)/f(b)$—hence equation (9.1) in the text.

In practice, we try to make $f(a)$ and $f(b)$ of opposite sign, to simplify (9.1); this generally gives better answers anyway. You must get accustomed to using the secant method—it really is useful.

9.4 ITERATION

Sometimes it is even easier to iterate for an answer. Consider the equation $\tan \theta = \theta$ (in radians). Since in the first positive branch, except for $\theta = 0$, $\tan \theta > \theta$, so there is no root below π. Above π, we have a root less than $3/2\pi$. So we 'iterate' on

$$\theta^* = \tan^{-1} \theta + \pi.$$

For example, $\theta = 3$ gives $\theta^* = 4.39$, then $\theta^* = 4.39$ gives $\theta^{**} = 4.49$, simply by leaving θ^* in the machine. It seems to be converging to something. Thus $3 \to 4.39 \to 4.49318 \to 4.49340 \to 4.49341$ which, if you try it, satisfies the equation! We can get the higher roots in the same way. It may take you a little time to find the right technique—for example if you iterate in the opposite way, it diverges—try $\theta^* = \tan^{-1}\theta - \pi$. Sometimes it works—the errors get progressively smaller and smaller; sometimes it doesn't. Try it and see.

9.5 ACCELERATED ITERATION

Long ago, when physical labour (turning a handle) was involved in iteration, the Aitken δ^2 technique was popular. Occasionally it is still useful today. Assume $a \to b \to c \to d$.... At any stage we can 'accelerate' the convergence and start again. A relatively 'clever' way of reducing the arithmetic—which even today you may occasionally find useful. If the iterated values, $a \to b \to c \cdots \to z$ (exact) form an exact geometric progression, i.e. if the error that comes out is exactly proportional to the error we put in, so that the error decreases by the same ratio at each iteration, then we can predict where they are going:

$$\frac{b - z}{a - z} = \frac{c - z}{b - z}$$

$$z = \frac{ac - b^2}{a + c - 2b}. \tag{9.2}$$

Note, $a + c - 2b$ is a sort of second difference, hence the name δ^2. Also $ac - b^2$ is parallel, term by term, with the denominator, which helps you to remember formula (9.2). It can even give convergence, when without it there would be mild divergence! (i.e. with (9.2) you can iterate the 'wrong' way if you are lucky.)

Another form exists. If following b you have a sudden inspiration, and you iterate with an unrelated number, $p \to q$,

$$z = \frac{aq - bp}{(a + q) - (b + p)}. \tag{9.3}$$

This is also easy to remember. Let us see the more general version working. Thus in solving $\tan\theta = \theta$, $4 \to 4.46741$ and $4.7 \to 4.50275$, hence $z = 4.49226$ by (9.3), which is almost right. There are alternative versions of (9.2) and (9.3) which generate less roundoff, but they are harder to remember. In the event of roundoff trouble, we would simply accelerate using only the last few digits, for example in this case $0 \to 0.46741$ and $0.7 \to 0.50275$, although in fact roundoff is not a problem, here.

9.6 SEVERAL VARIABLES

Experience has taught us that when there are several nonlinear equations to be solved, one should not automatically panic. If, for example, we have three equations, one of which contains only two variables, it is usually possible to get the roots by iterating: by solving that equation for one of the two variables, then solving the other two equations in turn for the other two variables; then starting again. This case is quite common, and is sufficient reason for your learning iteration.

Example.

Consider the equations in x, y, and z.

$$\tan^{-1} y = \sin x + \cos z \qquad \text{I}$$

$$z^x = \frac{1}{y} \qquad \text{II}$$

$$e^x = z^2 \qquad \text{III}$$

In general this sort of thing is difficult. The easiest way to start would be a massive search on the computer. Generate values of the errors in I, II, and III using three DO-loops in the x, y, and z, then find the vicinity of the roots by looking at the results. Complete the job by Newton–Raphson. (See below, section 9.7). Not at all pleasant.

But quite often, and in this case by design, you can use the direct approach:

(a)   Guess x, solve III for z, and then knowing x and z, solve II for y. Then substitute in I. Treat the error as if you were solving a single nonlinear equation in x. Otherwise,

(b)   Iterate thus: Guess x, then

$$\text{III} \rightarrow z = \sqrt{e^x}$$

$$\text{II} \rightarrow y = z^{-x}$$

$$\text{I} \rightarrow x = \sin^{-1}\{\tan^{-1} y - \cos z\}$$

In either case, the answers quickly settle down:

$$x = 0.425240055$$

$$y = 0.913552392$$

$$z = 1.236914568.$$

9.7   THE TANGENT METHOD: NEWTON–RAPHSON

First method last. It's not much use in hand calculations, but in a sense, it is what you would try first, if you didn't know any of those techniques (see Fig. 9.4). For example, maybe you want to compute $10^{2/3}$ on your kid sister's calculator, which has none of the usual functions? i.e., solve the equation:

$$x^3 - 100 = e(x) = 0$$

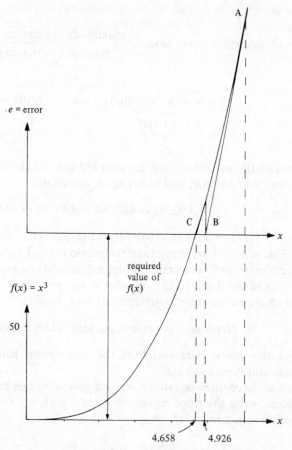

Fig. 9.4   Starting at A, the first guess of x, we proceed along the tangent to B, where the error should be zero. We are disappointed, however, because we are dealing with a curve, and the terms of the Taylor series above linear cannot be neglected. We now calculate e for the value of x at B, and we continue as if this is the first guess.

Guess an answer, say $x = 6$, and try to improve it.

$$e(6) = 116$$

$$\frac{de}{dx} = 3x^2 = 108 \text{ for } x = 6$$

'Linearizing' the function $e(x)$ means going straight along the tangent, with this slope. Mathematically this means that we neglect the higher terms in the Taylor series. If our improved answer is x^*,

$$e(x^*) \cong e(x) + (x^* - x)\frac{de(x)}{dx} \text{ by Taylor's series}$$

$$= 0 \text{ hopefully.}$$

$$x^* \cong x - e(x) \Big/ \frac{de(x)}{dx},$$

that is,

$$\text{new value} = \text{old value} - \frac{\text{magnitude of function at old value}}{\text{slope of function at old value}}.$$

Here,

$$x^* = x - (x^3 - 100)/3x^2 (= 6 - 116/108 = 4.926)$$

$$= \frac{2}{3}x + \frac{1}{3}\frac{100}{x^2}.$$

Program this as a 'black-box' on your kid sister's calculator, which turns one number into another, and run it again and again:

$$6 \rightarrow 4.92593 \rightarrow 4.65769 \rightarrow 4.64164 \rightarrow 4.64159$$

errors $1.35841 \rightarrow 0.28434 \rightarrow 0.01610 \rightarrow 0.00005 \rightarrow 0$.

This is more interesting than the previous black box. Somehow, it seems to accelerate itself. The first iteration reduces the error to 20%, the second to 5.7%, and the third to 0.3% of what it was previously. The phenomenon is known as 'second-order convergence': very roughly,

$$(\text{error after iteration}) \cong (\text{error before iteration})^2/7.$$

With this black box, therefore, the assumption leading to the Aitken accelerator does not hold.

If we have three equations, we can generalise this progressive 'updating' process: write the three equations as $e_i(x, y, z) = 0$, $i = 1 \ldots 3$; we want to improve x, y, z to x^*, y^*, z^*.

$$e_1 + (x^* - x)\frac{\partial e_1}{\partial x} + (y^* - y)\frac{\partial e_1}{\partial y} + (z^* - z)\frac{\partial e_1}{\partial z} = 0$$

$$e_2 + (x^* - x)\frac{\partial e_2}{\partial x} + (y^* - y)\frac{\partial e_2}{\partial y} + (z^* - z)\frac{\partial e_2}{\partial z} = 0$$

$$e_3 + (x^* - x)\frac{\partial e_3}{\partial x} + (y^* - y)\frac{\partial e_3}{\partial y} + (z^* - z)\frac{\partial e_3}{\partial z} = 0.$$

Thus with a good deal of effort before we start, we have given ourselves three simultaneous linear equations to solve.

For an example of this technique, we consider the two non-reducible equations

$$x^2 + \ln x = \cos y + \sin y$$

$$1/x + e^x = \tan y + \sec y$$

which we write

$$x^2 + \ln x - \cos y - \sin y = g(x, y) = 0$$

$$1/x + e^x - \tan y - \sec y = h(x, y) = 0.$$

We search for a root, which we find in the neighbourhood of $x = 1.1$, $y = 1.1$.
We tabulate some values in that vicinity

	$g(x, y)$			$h(x, y)$		
	$x = 1.0$	$x = 1.1$	$x = 1.2$	$x = 1.0$	$x = 1.1$	$x = 1.2$
$y = 1.0$		-0.0765			0.5050	
$y = 1.1$	-0.3448	-0.0395	0.2775	-0.4511	-0.2561	-0.0159
$y = 1.2$		0.0109			-1.4186	

From this table, we deduce for example that

$$\frac{\partial g(1.1, 1.1)}{\partial x} = \frac{0.2775 - (-0.3448)}{0.2}$$

$$= 3.1116 \text{ (carrying a few extra decimal places)}$$

Hence,

$$\begin{bmatrix} \partial g/\partial x & \partial g/\partial y \\ \partial h/\partial x & \partial h/\partial y \end{bmatrix} = \begin{bmatrix} 3.1116 & 0.4369 \\ 2.1758 & -9.6182 \end{bmatrix} = [J]$$

and

$$\begin{bmatrix} \partial g/\partial x & \partial g/\partial y \\ \partial h/\partial x & \partial h/\partial y \end{bmatrix}^{-1} = \begin{bmatrix} 0.3115 & 0.0141 \\ 0.0705 & -0.1008 \end{bmatrix} = [J]^{-1}$$

We could differentiate analytically at this stage, but it would probably take longer to do.

This matrix inversion anticipates the next chapter, it is really a means of solving the equations

$$[J] \begin{Bmatrix} \delta x \\ \delta y \end{Bmatrix} = - \begin{Bmatrix} -0.0395 \\ -0.2561 \end{Bmatrix} = - \begin{Bmatrix} g(x, y) \\ h(x, y) \end{Bmatrix}.$$

That is, the slopes times the differences in the current x and y from those at the roots, will give us the values of g and h at the current x and y.

We invert this equation to give

$$\begin{Bmatrix} \delta x \\ \delta y \end{Bmatrix} = - [J]^{-1} \begin{Bmatrix} g(x, y) \\ h(x, y) \end{Bmatrix}.$$

We thus calculate the changes in x and y for the 'new' values. Thus here, $[\delta x, \delta y] = [0.0159, -0.0230]$ and the new $[x, y]$ becomes $[1.1159, 1.0770]$ giving $g = 0.0004$, $h = -0.0192$. We can now assume that $[J]^{-1}$ is unchanged, giving $\delta x = 0.0002$, $\delta y = -0.0020$, so that $x = 1.1161$, $y = 1.0750$ giving $g = 0.0002$, $h = -0.0020$. Again, $[J]^{-1}$ gives $\delta x = -0.00003$, $\delta y = -0.00022$; another iteration with $y = 1.0768$ reduced δy to -0.00005, so at this stage we probably have the correct answers to four places. (We could work to 10 decimals without calculating $[J]^{-1}$ again.)

We have kept the example simple, by having only two variables. You probably agree, it is hard work!

EXERCISES

9.1. The infinite series

$$\ln 2 = 1 - \tfrac{1}{2} + \tfrac{1}{3} - \tfrac{1}{4} + \tfrac{1}{5} \cdots$$

converges slowly. The sequence of sums oscillates:

No of terms (n)	Sum of terms (S_n)
1	1
2	0.5
3	0.8333.
4	0.78333.
5	0.61666.
6	0.759523803

Apply the Aitken δ^2 technique (formula (9.2)) to S_3, S_4, and S_5 (as a, b, and c), then to S_4, S_5, and S_6, then to S_5, S_6, and S_7. Apply the formula to the three numbers which result (note they also oscillate). Compare your final answer with the exact value for ln 2.

9.2. Compute the root of $2x^3 - 3x - 1 = 0$ in the region $1 \le x \le 2$ by Newton's method and by the secant method. Use $x = 1$ and $x = 2$ as the starting values of the secant method, and compare the answers with those from Newton's method after 5 iterations starting from $x = 1$ and $x = 2$.

9.3. In open channel flow, water will flow at a critical depth y_{cr} when the Froude Number equals 1.0. The Froude Number F_r is given by

$$F_r^2 = \frac{Q^2 B}{g A^3}$$

where Q is the flow rate (volume of water/second);
     B is the surface width of the water in the channel;
     A is the cross-sectional area of the channel;
     g is the gravitational constant.

What is the critical depth of water in a channel which is trapezoidal in cross-section with a base width of 6 m and equal side slopes of 1.5 horizontal: 1.0 vertical, when the flow rate is 25 m³/s? See figure.

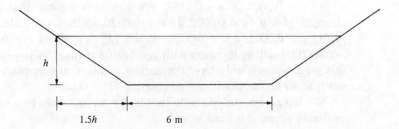

h

1.5h 6 m

9.4. Find as many roots as possible as accurately as possible in 20 minutes for

$$\tan x = z/y$$

$$y = xe^z$$

$$z = \ln(xy - 1)$$

Hint: first simplify the equation.

9.5. For the two non-reducible equations

$$e^x + 8 \sin x = 16 - 2 \ln y$$

$$8e^{-x} - 3 = \cos y - 5 \sin y$$

It is noted that there is a root in the neighbourhood of $x = 1.8$, $y = 3.0$. Use our technique of section 9.7 to find the roots. It ought to help that

$$\begin{vmatrix} 4.24514634 & 0.66691375 \\ -1.32459619 & -4.80083174 \end{vmatrix}^{-1} = \begin{vmatrix} 0.24623638 & 0.03420624 \\ -0.06793901 & -0.21773508 \end{vmatrix}$$

9.6. Water flows out of a reservoir down a channel of rectangular cross-section of width 12 m, at a rate of $Q = 300$ m^3/s. The water initially flows supercritically and then passes through a hydraulic jump to subcritical flow, as depicted.

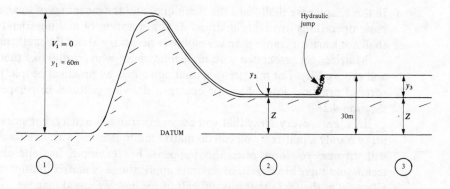

Conservation of energy between points 1 and 2 gives

$$\text{total head } H = 60 \text{ m} = Z + y_2 + \frac{V_2^2}{2g} \quad (V_2 = \text{velocity at position 2})$$

Conservation of momentum between points 2 and 3 gives

$$\frac{q^2}{g} = \tfrac{1}{2} y_2 y_3 (y_2 + y_3)$$

where q = volume flow rate/unit width of channel. Determine y_2, y_3, and Z as depicted. (Note $y_3 = 30 - Z$, and $q = y_2 V_2$ m^2/s).

10

Matrices

10.1 WHO USES THEM?

In this chapter we shall learn the techniques and the meanings of some of the basic operations involving matrices. It is a question of making friends. We shall not learn any more than is absolutely necessary about the mathematics.

Matrices get associated with mathematicians, who (really) like theorems and proofs. Pity. For matrices are used nowadays by practical people, for all sorts of practical jobs—like, for example, drawing pictures in perspective! (Section 4.2).

In a sense, every array that you use in Fortran is a matrix; but strictly, an array is only a matrix if you can do matrix multiplication with it. However, it will surprise you how often this happens. For example, in finite element techniques there are dozens of separate applications of matrix multiplication, all useful in the sense that it is difficult to see how we could manage without them. In electrical circuit theory the use of matrices is not a recent phenomenon. However, finite elements have (or perhaps do not have) the present record, with matrices in the order of a million × million. (This was a sparse matrix. In other words, there were certainly not 10^{12} non-zero terms: most were zero.)

But Fortran† is an unfortunate name for a programming language. You should try *not* to code big formulae. It isn't clever. It leads to nasty mistakes, obscure and difficult to find. Much better, find out how that complicated formula was developed and code it accordingly. That puts some engineering sense into the process of coding. Probably the ultimate in breaking down formulae into their simplest terms is to use matrices.

---

† Acronym for 'formula translations'.

Yet, matrix algebra is a traditionally distasteful subject to engineers, much more so than vector algebra, which we can usually understand physically. Pity. Matrices are a very natural way to use a computer, especially a supercomputer. Besides, we saw in section 7.2 how simply writing something in matrix form occasionally clarifies the theory behind what is really happening.

However, in the real world outside it is unlikely that you will ever have to perform matrix arithmetic to any extent, much less that you will ever have to code anything. Matrices are very widely used in industry, but 'black boxes', i.e. public subroutines, are already very common. The useful worker in today's world is somebody who understands the lingo, who knows what is happening, and especially who knows what can go wrong. 'Matrix literacy'. You must have it.

For the moment, the best entry to matrices is through equation solving. As at the end of the previous chapter. This is why we write the matrix operations alongside the familiar manipulations of the equations.

10.2   THE GAUSS–JORDAN SOLUTION

To review matrix multiplication, we write three very simple simultaneous equations, and we write the same operations in matrix notation:

$$
\begin{array}{ll}
\text{I.} & x - y + z = 2 \\
\text{II.} & -x + 5y + 3z = 18 \\
\text{III.} & x + 3y + 14z = 49
\end{array}
\qquad
\begin{bmatrix} 1 & -1 & 1 \\ -1 & 5 & 3 \\ 1 & 3 & 14 \end{bmatrix}
\begin{Bmatrix} x \\ y \\ z \end{Bmatrix}
=
\begin{Bmatrix} 2 \\ 18 \\ 49 \end{Bmatrix}
$$

First we add equation I to equation II, and then subtract equation I from equation III, giving us:

$$x - y + z = 2$$

$$4y + 4z = 20$$

$$4y + 13z = 47.$$

In matrix form:

$$
\begin{bmatrix} 1 & 0 & 0 \\ 1 & 1 & 0 \\ -1 & 0 & 1 \end{bmatrix}
\begin{bmatrix} 1 & -1 & 1 & 2 \\ -1 & 5 & 3 & 18 \\ 1 & 3 & 14 & 49 \end{bmatrix}
=
\begin{bmatrix} 1 & -1 & 1 & 2 \\ 0 & 4 & 4 & 20 \\ 0 & 4 & 13 & 47 \end{bmatrix}
$$

(The 4 × 3 matrices here comprise the coefficients of the left-hand side, and the results in the right-hand side combined.) The process of eliminating x from equations II and III is exactly equivalent to pre-multiplying by the 3 × 3 matrix above, containing several zeros. In the top row of the 3 × 3, we are just multiplying equation I by 1: in the second row we are adding one equation I to one equation II, etc. Observe how in the resultant matrix the coefficient of x is now zero in rows II and III. In effect, we have 'substituted' for x from I, into II and III. Now to eliminate y from equations I and III, we add $\frac{1}{4}$ of II to I, and subtract II from III, giving:

$$x \quad + 2z = 7$$
$$4y + 4z = 20$$
$$9z = 27$$

$$\begin{bmatrix} 1 & \frac{1}{4} & 0 \\ 0 & 1 & 0 \\ 0 & -1 & 1 \end{bmatrix} \begin{bmatrix} 1 & -1 & 1 & 2 \\ 0 & 4 & 4 & 20 \\ 0 & 4 & 13 & 47 \end{bmatrix} = \begin{bmatrix} 1 & 0 & 2 & 7 \\ 0 & 4 & 4 & 20 \\ 0 & 0 & 9 & 27 \end{bmatrix}$$

Again, the matrix operation gives the same numbers. Finally we subtract 2/9 of III from I, and 4/9 of III from II:

$$x \quad\quad = 1$$
$$4y \quad = 8$$
$$9z = 27$$

$$\begin{bmatrix} 1 & 0 & -\frac{2}{9} \\ 0 & 1 & -\frac{4}{9} \\ 0 & 0 & 1 \end{bmatrix} \begin{bmatrix} 1 & 0 & 2 & 7 \\ 0 & 4 & 4 & 20 \\ 0 & 0 & 9 & 27 \end{bmatrix} = \begin{bmatrix} 1 & 0 & 0 & 1 \\ 0 & 4 & 0 & 8 \\ 0 & 0 & 9 & 27 \end{bmatrix}.$$

This is easily solved. But let's take the formal approach one stage further: dividing II by 4 and III by 9,

$$x \quad\quad = 1$$
$$y \quad = 2$$
$$z = 3$$

$$\begin{bmatrix} 1 & 0 & 0 \\ 0 & \frac{1}{4} & 0 \\ 0 & 0 & \frac{1}{9} \end{bmatrix} \begin{bmatrix} 1 & 0 & 0 & 1 \\ 0 & 4 & 0 & 8 \\ 0 & 0 & 9 & 27 \end{bmatrix} = \begin{bmatrix} 1 & 0 & 0 & 1 \\ 0 & 1 & 0 & 2 \\ 0 & 0 & 1 & 3 \end{bmatrix}.$$

These are the explicit answers. Altogether we have pre-multiplied our starting matrix [A] by four simple matrices in turn.

$$\begin{bmatrix} 1 & 0 & 0 \\ 0 & \frac{1}{4} & 0 \\ 0 & 0 & \frac{1}{9} \end{bmatrix} \begin{bmatrix} 1 & 0 & -\frac{2}{9} \\ 0 & 1 & -\frac{4}{9} \\ 0 & 0 & 1 \end{bmatrix} \begin{bmatrix} 1 & \frac{1}{4} & 0 \\ 0 & 1 & 0 \\ 0 & -1 & 1 \end{bmatrix} \begin{bmatrix} 1 & 0 & 0 \\ 1 & 1 & 0 \\ -1 & 0 & 1 \end{bmatrix} [A]$$

$$= \frac{1}{36} \begin{bmatrix} 61 & 17 & -8 \\ 17 & 13 & -4 \\ -8 & -4 & 4 \end{bmatrix} [A].$$

The combined pre-matrix is known as the 'inverse matrix' as in section 9.7; but this was not a good way to find $[A]^{-1}$—more later. If you want to check the arithmetic, the easiest way is to multiply the first two matrices, and the last two, and then multiply the resulting matrices together. With matrices, unlike vectors, brackets make no difference: $\mathbf{ABCD} = (\mathbf{AB})(\mathbf{CD}) = (\mathbf{ABC})\mathbf{D} = \mathbf{A}(\mathbf{BCD})$, etc. But never change the order, unless you know what you are doing.

It is not even a good way to solve equations, but it did introduce some interesting ideas:

(i) If the coefficient of x in I had not been unity, we should have had to divide by it. With N equations in N variables, there are exactly N numbers that we have to divide by, in this case 1, 4, and 9. These are known as the three 'pivots'.

(ii) Calculate the determinant:

$$\begin{vmatrix} 1 & -1 & 1 \\ -1 & 5 & 3 \\ 1 & 3 & 14 \end{vmatrix} = (1)\begin{vmatrix} 5 & 3 \\ 3 & 14 \end{vmatrix} - (-1)\begin{vmatrix} -1 & 3 \\ 1 & 14 \end{vmatrix} + (1)\begin{vmatrix} -1 & 5 \\ 1 & 3 \end{vmatrix}$$

$$= 1(5 \times 14 - 3 \times 3) + 1(-1 \times 14 - 3 \times 1)$$

$$+ 1(-1 \times 3 - 5 \times 1)$$

$$= 36.$$

This is the product of the pivots, $1 \times 4 \times 9$. The sign is correct, because we took the pivots in the natural order. Whenever we switch two equations, we reverse the sign.

(iii) The pivots are all positive. In this case, we say that the matrix is 'positive definite'. Important—much more later.

The Gauss–Jordan method is seldom used; it is often too costly. But it is very easy to code. Hardly worth mentioning today, with 'black boxes' everywhere!

10.3 GAUSS REDUCTION AND BACK SUBSTITUTION

Most people nowadays use a slightly more sophisticated technique to solve equations.

$$x - y + z = 2 \quad (3)$$

$$-x + 5y + 3z = 18 \quad (25)$$

$$x + 3y + 14z = 49 \quad (67)$$

Each number in brackets is the sum of the coefficients and the right-hand side, e.g. $(67) = 1 + 3 + 14 + 49$. Such numbers will provide a useful running check later. We start as before:

$$\begin{array}{ll} x - y + z = 2 & (3) \\ 4y + 4z = 20 & (28) \\ 4y + 13z = 47 & (64) \end{array} \qquad \begin{bmatrix} 1 & 0 & 0 \\ 1 & 1 & 0 \\ -1 & 0 & 1 \end{bmatrix}\begin{bmatrix} 1 & -1 & 1 & 2 & 3 \\ -1 & 5 & 3 & 18 & 25 \\ 1 & 3 & 14 & 49 & 67 \end{bmatrix}.$$

The bracketed numbers are treated as additional right-hand sides. We now eliminate y from equation III by subtracting II from III:

$$x - y + z = 2 \quad (3)$$
$$4y + 4z = 20 \quad (28)$$
$$9z = 27 \quad (36)$$

$$\begin{bmatrix} 1 & 0 & 0 \\ 0 & 1 & 0 \\ 0 & -1 & 1 \end{bmatrix} \begin{bmatrix} 1 & -1 & 1 & 2 & 3 \\ 0 & 4 & 4 & 20 & 28 \\ 0 & 4 & 13 & 47 & 64 \end{bmatrix}.$$

Whenever we complete a line, we check that the bracketed number is still the sum of the other numbers on that line. Summarising these two steps:

$$\begin{bmatrix} 1 & 0 & 0 \\ 0 & 1 & 0 \\ 0 & -1 & 1 \end{bmatrix} \begin{bmatrix} 1 & 0 & 0 \\ 1 & 1 & 0 \\ -1 & 0 & 1 \end{bmatrix} \begin{bmatrix} 1 & -1 & 1 & 2 & 3 \\ -1 & 5 & 3 & 18 & 25 \\ 1 & 3 & 14 & 49 & 67 \end{bmatrix}$$

$$= \begin{bmatrix} 1 & 0 & 0 \\ 1 & 1 & 0 \\ -2 & -1 & 1 \end{bmatrix} \begin{bmatrix} 1 & -1 & 1 & 2 & 3 \\ -1 & 5 & 3 & 18 & 25 \\ 1 & 3 & 14 & 49 & 67 \end{bmatrix}$$

$$= \begin{bmatrix} 1 & -1 & 1 & 2 & 3 \\ 0 & 4 & 4 & 20 & 28 \\ 0 & 0 & 9 & 27 & 36 \end{bmatrix} \tag{10.1}$$

Before we carry on to the solution, some definitions: matrix literacy.

Definition:
'Diagonal'—the Nth terms on the Nth lines of a matrix.

Definition:
'Lower Triangular matrix'—one whose terms above the diagonals are zero, with some (or all) of those below the diagonal being non-zero.

Definition:
'Upper Triangular matrix'—one whose terms below the diagonals are zero.

Thus in Gauss reduction, we transform the left-hand side matrix into an upper triangular matrix. We do this, by pre-multiplying by 'lower triangles', whose diagonal terms are unity. The product of such matrices is also a lower triangle with unit diagonals, as above. Let's now complete the solution:

$$x - y + z = 2 \quad (3)$$
$$4y + 4z = 20 \quad (28)$$
$$9z = 27 \quad (36).$$

In the 'Back Substitution' we use III first:

$$\text{From III, } z = 3 \quad (Z = 4)$$
$$\text{From II, } y = 2 \quad (Y = 3)$$
$$\text{From I, } x = 1 \quad (X = 2)$$

For the check solutions, Z, Y, and X, we treat (3), (28), and (36) as the right-hand sides. X, Y, and Z should exceed x, y, and z by 1; okay—check completed. The method is easily programmed: see our subroutine GAUSS after programs LEASQ and CURFIT.

10.4   INTO BATTLE: ROUNDOFF

Let's have a taste of the real world—genuine numbers now. Keep only 6 decimals:

$99x + 106y + 77z = 275$ (557)** (an equation to be back-substituted)

$106x + 114y + 87z = 299$ (606)

$77x + 87y + 101z = 255$ (520)

II → II − 1.070707 I, III → III − 0.777778 I

$\qquad 0.505058y + 4.555561z = 4.555575 \quad (9.616201)$**

$\qquad 4.555532y + 41.111094z = 41.111050 \quad (86.777654)$

III → III − 9.019820 II

$\qquad\qquad 0.020754z = 0.020584 \ (0.041252)$**

Back-substituting in the asterisked equations:

$$z = 0.991809 \ (Z = 1.987665)$$

$$y = 0.073910 \ (Y = 1.111302)$$

$$x = 1.927235 \ (X = 2.890422).$$

The 'correct' answers are $x = 2$, $y = 0$, $z = 1$, yet substituting these appalling answers into the original equations gives (275.000018, 299.000033, 254.999974) instead of (275,299,255). It would appear that all is well!

There is some indication that all is not well, however, in the 'check' values X, Y, and Z, but the situation is much worse even than these disagreements suggest. This is typical. The only reliable check is to estimate the correction, by the 'tangent' method (section 9.7):

$$99(x^* − x) + 106(y^* − y) + 77(z^* − z) = −0.000018$$

$$106(x^* − x) + 114(y^* − y) + 87(z^* − z) = −0.000033$$

$$77(x^* − x) + 87(y^* − y) + 101(z^* − z) = 0.000026.$$

where the new right-hand sides are the errors in the check substitutions; *not*, and observe this carefully, with the intention of improving the answers to (x^*, y^*, z^*) but merely to discover the order of magnitude of the errors. Nevertheless, in this particular case the corrections will be quite good. Try it.

10.5 MATRIX TERMINOLOGY

Definition:
'Ill-conditioned matrix'—one that is abnormally sensitive to roundoff errors.

Definition:
'Singular matrix'—at least one row is a linear combination of the others, therefore in general the equations cannot be solved.

Definition:
A 'Transposed' matrix,

$$\begin{bmatrix} 1 & 2 \\ 3 & 4 \end{bmatrix}^{T} = \begin{bmatrix} 1 & 3 \\ 2 & 4 \end{bmatrix}$$

is reflected about the diagonals. Column N becomes row N, and vice versa. Observe,

$$\left\{ \begin{bmatrix} 1 & 2 \\ 5 & 6 \end{bmatrix} \begin{bmatrix} 3 & 7 \\ 4 & 8 \end{bmatrix} \right\}^{T} = \begin{bmatrix} 11 & 23 \\ 39 & 83 \end{bmatrix}^{T} = \begin{bmatrix} 11 & 39 \\ 23 & 83 \end{bmatrix}$$

$$= \begin{bmatrix} 3 & 4 \\ 7 & 8 \end{bmatrix} \begin{bmatrix} 1 & 5 \\ 2 & 6 \end{bmatrix} = \begin{bmatrix} 3 & 7 \\ 4 & 8 \end{bmatrix}^{T} \begin{bmatrix} 1 & 2 \\ 5 & 6 \end{bmatrix}^{T}$$

or, in general $[\mathbf{AB}]^{T} = \mathbf{B}^{T}\mathbf{A}^{T}$. Invent two new matrices, and check this again for yourself.

If $\mathbf{A} = \mathbf{A}^{T}$, then $\mathbf{A}$ must be symmetric.

$$\det \mathbf{A} = \det \mathbf{A}^{T}$$

The determinant of a product equals the product of the determinants:

$$\det[\mathbf{AB}] = \det[\mathbf{A}]\det[\mathbf{B}]$$

10.6 INVERTING AN UNSYMMETRICAL MATRIX; CHOOSING PIVOTS

Until now, we have taken the first diagonal as the first pivot, then the second as the second, etc. We were allowed to do this only because our matrices were (a) symmetric, and (b) positive definite. (We will define 'positive definite' in section 11.7, when you have learned more.) Suppose for example the first diagonal had been zero!

Two new developments. Our matrix is now unsymmetrical, and although Gauss's method sometimes works, we solve the equations in a different way, so as to create the inverse matrix by a technique commonly used by programmers.

$$\begin{array}{rl} \text{I} & x - 2y + 3z = a \\ \text{II} & 2x - 5y + 5z = b \\ \text{III} & 2y + z = c. \end{array}$$

Note, a, b, and c are also variables. We choose the largest coefficient, 5, as the first pivot. (-5 would have been equally good.) We solve II for z, $-\frac{2}{5}x + y + \frac{1}{5}b = z$ and we substitute z into the other equations.

$$-\frac{1}{5}x + y + \frac{3}{5}b = a$$

$$-\frac{2}{5}x + y + \frac{1}{5}b = z$$

$$-\frac{2}{5}x + 3y + \frac{1}{5}b = c.$$

The left-hand variables are now (x, y, b) and the right-hand variables are (a, z, c). We have exchanged z with b. Once again, we find the largest pivot in the remaining equations, namely 3. We solve III for y, and substitute y into I and II:

$$-\frac{1}{15}x + \frac{1}{3}c + \frac{8}{15}b = a$$

$$-\frac{4}{15}x + \frac{1}{3}c + \frac{2}{15}b = z$$

$$\frac{2}{15}x + \frac{1}{3}c - \frac{1}{15}b = y.$$

There is now only one choice; to solve I for x and to substitute in II and III. The pivot will be small, but never arbitrarily small, because the product of the pivots is still the determinant, apart from the sign. If at any stage it is not possible to find a non-zero pivot, then (a) the pivot is zero, which implies (b) the determinant is zero, so (c) the matrix can't be inverted, and (d) the matrix is called 'singular'. Proceeding with the inversion:

$$-15a + 5c + 8b = x$$

$$4a - c - 2b = z$$

$$-2a + c + b = y.$$

We have a, b, c on the left-hand side, x, y, z on the right.

Let's now put a, b, c and x, y, z into alphabetical order.

$$-15a + 8b + 5c = x$$

$$-2a + b + c = y$$

$$4a - 2b - c = z.$$

The standard check that we have the inverse matrix is to multiply this into the matrix we started with.

$$\begin{bmatrix} 1 & -2 & 3 \\ 2 & -5 & 5 \\ 0 & 2 & 1 \end{bmatrix} \begin{bmatrix} -15 & 8 & 5 \\ -2 & 1 & 1 \\ 4 & -2 & -1 \end{bmatrix} = \begin{bmatrix} 1 & 0 & 0 \\ 0 & 1 & 0 \\ 0 & 0 & 1 \end{bmatrix} = [I].$$

This process, or something like it, forms the basis of most matrix inversion subroutines. The next section discusses $[I]$, and the algebra of matrix inversion.

10.7 MORE MATRIX DEFINITIONS

Definition:
[I] is a unit 'diagonal matrix' (all non-diagonal terms zero.) It is also called
the 'identity matrix' because anything multiplied by it is unchanged:

$$IA = AI = A$$

Definition:
If $AB = I$, then $B = A^{-1}$ is termed the 'inverse' of A.
The pre-inverse and the post-inverse are equal. Let

$$AB = I, CA = I.$$

Then

$$CAB = C(AB) = CI = C$$

$$= (CA)B = IB = B.$$

Thus $B = C$. (This is really not a trivial conclusion!) The inverse of a product:

$$[FG]^{-1} = G^{-1}F^{-1}.$$

For

$$[FG][G^{-1}F^{-1}] = F(GG^{-1})F^{-1}$$

$$= F[I]F^{-1}$$

$$= FF^{-1} = I.$$

In transposing a product, $(FG)^T = I^T = I = G^TF^T$ from section 10.5; it
follows that $(F^T)^{-1} = (F^{-1})^T$. Hence the inverse of a symmetric matrix is also
symmetric.

The determinant of the inverse is the inverse of the determinant; if $AB = I$,
then

$$\det A \det B = \det I = 1.$$

10.8 SYMMETRIC PART-INVERSION

If your matrix is positive definite, you can use this cheaper variant.

$$10w + 4x + 6y + 3z = a$$

$$4w + 9x + 2y - z = b$$

$$6w + 2x + 8y + z = c$$

$$3w - x + y + 7z = d.$$

For illustration, we make only the first part-inversion.

$$-0.1a + 0.4x + 0.6y + 0.3z = -w$$

$$0.4a + 7.4x - 0.4y - 2.2z = b$$

$$0.6a - 0.4x + 4.4y - 0.8z = c$$

$$0.3a - 2.2x - 0.8y + 6.1z = d$$

$$\begin{bmatrix} -0.1 & 0.4 & 0.6 & 0.3 \\ 0.4 & 7.4 & -0.4 & -2.2 \\ 0.6 & -0.4 & 4.4 & -0.8 \\ 0.3 & -2.2 & -0.8 & 6.1 \end{bmatrix} \begin{Bmatrix} a \\ x \\ y \\ z \end{Bmatrix} = \begin{Bmatrix} -w \\ b \\ c \\ d \end{Bmatrix}.$$

We solved for w in equation I, but then we changed the sign of w. With this trivial change, we preserve symmetry. This is worthwhile, because with symmetry we halve the arithmetic. A professional programmer would reduce the storage by about half too, by only storing the diagonal terms and above. Admittedly, with the sign change, the inverse matrix turns out to be negative, but we can live with that. We use this method in our program POSIN.

10.9   TRIPLE MATRIX PRODUCTS BY PART-REDUCTION

Occasionally we need $\mathbf{AB}^{-1}\mathbf{C}$. More often, we need $\mathbf{G}^T\mathbf{F}^{-1}\mathbf{G}$, where $\mathbf{F}$ is symmetric positive definite (we leave the definition of positive definite to section 11.7). For example, consider

$$\mathbf{F}p + \mathbf{G}y = ?$$

$$\mathbf{G}^T p = ??$$

Thus $p = \mathbf{F}^{-1}[? - \mathbf{G}y]$, and the second block of equations becomes

$$-\mathbf{G}^T\mathbf{F}^{-1}\mathbf{G}y = ???$$

which gives the solution. The question marks show that we are not interested in the values at this stage.

To show this numerically, consider a pair of equations like those above.

$$\begin{bmatrix} 1 & 1 & \vdots & 3 & 5 \\ 1 & 2 & \vdots & 4 & 6 \\ \cdots & \cdots & \cdots & \cdots & \cdots \\ 3 & 4 & \vdots & 0 & 0 \\ 5 & 6 & \vdots & 0 & 0 \end{bmatrix} \begin{Bmatrix} p \\ q \\ y \\ z \end{Bmatrix} = \begin{Bmatrix} ? \\ ? \\ ? \\ ? \end{Bmatrix}.$$

The matrix is 'partitioned' by the dotted lines. We do not mean by the question marks that all the right-hand sides are the same: as before we are not interested in the values; p, q, y, z are 'dummy variables'. We eliminate p and q in turn, and then we stop: II → II − I, III → III − 3I, IV → IV − 5I

$$\left[\begin{array}{cc:cc} 1 & 1 & 3 & 5 \\ 0 & 1 & 1 & 1 \\ \hdashline 0 & 1 & -9 & -15 \\ 0 & 1 & -15 & -25 \end{array}\right] \left\{\begin{array}{c} p \\ q \\ y \\ z \end{array}\right\} = \left\{\begin{array}{c} ?? \\ ?? \\ ?? \\ ?? \end{array}\right\}.$$

Now III → III − II and IV → IV − II

$$\left[\begin{array}{cc:cc} 1 & 1 & 3 & 5 \\ 0 & 1 & 1 & 1 \\ \hdashline 0 & 0 & -10 & -16 \\ 0 & 0 & -16 & -26 \end{array}\right] \left\{\begin{array}{c} p \\ q \\ y \\ z \end{array}\right\} = \left\{\begin{array}{c} ??? \\ ??? \\ ??? \\ ??? \end{array}\right\}.$$

Finally,

$$\left[\begin{array}{cc} -10 & -16 \\ -16 & -26 \end{array}\right] = -\left[\begin{array}{cc} 3 & 4 \\ 5 & 6 \end{array}\right]\left[\begin{array}{cc} 1 & 1 \\ 1 & 2 \end{array}\right]^{-1}\left[\begin{array}{cc} 3 & 5 \\ 4 & 6 \end{array}\right]$$

which is $-\mathbf{G}^{\mathrm{T}}\mathbf{F}^{-1}\mathbf{G}$, as required. The change in sign is avoided if we insert [F] negative. Isn't it strange, that the triple product involving the inversion is more direct than $\mathbf{G}^{\mathrm{T}}\mathbf{F}\mathbf{G}$?

10.10 THE CHOLESKI 'SQUARE ROOT': MORE WISDOM

We come finally to a technique which is only possible for a symmetric positive definite matrix. Writing

$$\left[\begin{array}{cccc} A & B & D & G \\ B & C & E & H \\ D & E & F & I \\ G & H & I & J \end{array}\right] = \left[\begin{array}{cccc} a & 0 & 0 & 0 \\ b & c & 0 & 0 \\ d & e & f & 0 \\ g & h & i & j \end{array}\right]\left[\begin{array}{cccc} a & b & d & g \\ 0 & c & e & h \\ 0 & 0 & f & i \\ 0 & 0 & 0 & j \end{array}\right]$$

we can solve for a, b, c... in that order

$$A = a^2 \qquad\qquad a = \sqrt{A}$$

$$B = ab \qquad\qquad b = B/a$$

$$C = b^2 + c^2 \qquad\qquad c = \sqrt{(C - b^2)}$$

$$D = ad \qquad\qquad d = D/a$$

$$E = bd + ce \qquad\qquad e = (E - bd)/c$$

$$F = d^2 + e^2 + f^2 \qquad\qquad f = \sqrt{(F - d^2 - e^2)}$$

$$G = ag \qquad\qquad g = G/a$$

$$H = bg + ch \qquad\qquad h = (H - bg)/c$$

$$I = dg + eh + fi \qquad\qquad i = (I - dg - eh)/f$$

$$J = g^2 + h^2 + i^2 + j^2 \qquad\qquad j = \sqrt{(J - g^2 - h^2 - i^2)}.$$

This is 'Choleski decomposition'. As in the preceding techniques, we do not need a second vector, [*abcd*...] in the computer, because we can replace *A* by *a*, *B* by *b*, etc., in turn—we never need *A* again. The [*abcd*...] form makes life easier later. Applying these formulae to the matrix of section 10.2,

$$
\begin{bmatrix} 1 & -1 & 1 \\ -1 & 5 & 3 \\ 1 & 3 & 14 \end{bmatrix} = \begin{bmatrix} 1 & 0 & 0 \\ -1 & 2 & 0 \\ 1 & 2 & 3 \end{bmatrix} \begin{bmatrix} 1 & -1 & 1 \\ 0 & 2 & 2 \\ 0 & 0 & 3 \end{bmatrix}
$$

by Choleski. Using Gauss reduction (Section 10.3), we then manipulate the Choleski factors above into lower (L_1) and upper (U_1) triangular matrices with unit diagonals.

The above

$$
= \begin{bmatrix} 1 & 0 & 0 \\ -1 & 1 & 0 \\ 1 & 1 & 1 \end{bmatrix} \begin{bmatrix} 1 & 0 & 0 \\ 0 & 2 & 0 \\ 0 & 0 & 3 \end{bmatrix} \begin{bmatrix} 1 & 0 & 0 \\ 0 & 2 & 0 \\ 0 & 0 & 3 \end{bmatrix} \begin{bmatrix} 1 & -1 & 1 \\ 0 & 1 & 1 \\ 0 & 0 & 1 \end{bmatrix}.
$$

By combining the two diagonal matrices into one, we obtain

$$
= \begin{bmatrix} 1 & 0 & 0 \\ -1 & 1 & 0 \\ 1 & 1 & 1 \end{bmatrix} \begin{bmatrix} 1 & 0 & 0 \\ 0 & 4 & 0 \\ 0 & 0 & 9 \end{bmatrix} \begin{bmatrix} 1 & -1 & 1 \\ 0 & 1 & 1 \\ 0 & 0 & 1 \end{bmatrix}
$$

$$
= \begin{bmatrix} 1 & 0 & 0 \\ -1 & 1 & 0 \\ 1 & 1 & 1 \end{bmatrix} \begin{bmatrix} 1 & -1 & 1 \\ 0 & 4 & 4 \\ 0 & 0 & 9 \end{bmatrix}.
$$

where the second matrix is what we got by complete Gauss reduction on the original matrix. In fact, this sequence is packed with wisdom. Refer back to sections 10.2 and 10.3. The diagonal matrix [1 4 9] contains the pivots: we call it **P**. Thus [1 2 3] is—it really is—$\sqrt{P}$. In the Gauss reduction, we operated with a lower triangle having unit diagonals. Observe that

$$
L_1 = \begin{bmatrix} 1 & 0 & 0 \\ -1 & 1 & 0 \\ 1 & 1 & 1 \end{bmatrix} = \begin{bmatrix} 1 & 0 & 0 \\ 1 & 1 & 0 \\ -2 & -1 & 1 \end{bmatrix}^{-1}.
$$

Indeed, L_1^{-1} is always another lower triangle, with unit diagonals. We have seen L_1^{-1} as the final Gauss reduction matrix in equation (10.1).

Summarising, the original matrix [**A**] may be written $A = L_1 P U_1$ where $U_1 = L_1^T$. Choleski becomes $A = (L_1\sqrt{P})(\sqrt{P}U_1)$. Gauss reduction gives $A^* = L_1^{-1}A = PU_1$. So they are in fact almost the same process. The moral is this: Choleski depends on A being positive definite. But one finds non-positive definite matrices for which diagonal pivoting is okay, provided the order is intelligently chosen. In fact, some people simply choose the largest diagonal pivot, positive or negative, and appear to thrive; but for example they would have trouble with

$$\begin{bmatrix} 0 & 1 \\ 1 & 0 \end{bmatrix}.$$

Gauss reduction allows them to do this. Choleski would not.

The form L_1PU_1 shows that the determinant is the product of the pivots, that is, the determinant of $[P]$, because the determinants of L_1 and U_1 are unity.

There is still much to learn about matrices, which is just as important in the world outside. Relax for now.

EXERCISES

10.1. Use Gauss elimination and backward substitution to solve

$$x + 2y + 3z = 20$$
$$w \quad\;\; + 3y + 2z = 18$$
$$2w + 3x \quad\;\; + z = 12$$
$$3w + 4x + y \quad\;\; = 14.$$

Hint: rearrange the equations first.

10.2. Solve the ill-conditioned equations retaining only 5 decimal place accuracy at each stage. Check by substituting answers.

$$52x + 67y + 46z = 175$$
$$67x + 73y + 54z = 194$$
$$46x + 54y + 65z = 165.$$

10.3. Invert

$$\begin{bmatrix} 2 & -3 & 6 \\ 2 & 4 & -5 \\ 2 & 6 & -3 \end{bmatrix}$$

using the method of section 10.3.

10.4. Study our program POSIN to see how to program symmetric part inversion, and then use the program to invert some matrices.

10.5. Use the Choleski square root method to write the matrix below in the form L_1PU_1.

$$\begin{bmatrix} 1 & -1 & 1 & 1 \\ -1 & 5 & -5 & 3 \\ 1 & -5 & 21 & -15 \\ 1 & 3 & -15 & 46 \end{bmatrix}$$

10.6. A resistive electric circuit has a 50 V source as shown. The equations for current at nodes 1-7 are (using Ohm's and Kirchhoff's Laws):

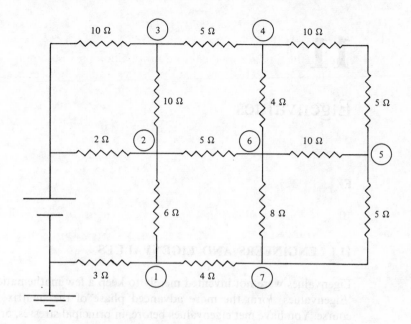

1. $[\frac{1}{3} + \frac{1}{6} + \frac{1}{4}]V_1 - \frac{1}{6}V_2 - \frac{1}{4}V_7 = 0$
2. $-\frac{1}{6}V_1 + [\frac{1}{2} + \frac{1}{10} + \frac{1}{5} + \frac{1}{6}]V_2 - \frac{1}{10}V_3 - \frac{1}{5}V_6 = 25$
3. $-\frac{1}{10}V_2 + [\frac{1}{10} + \frac{1}{10} + \frac{1}{5}]V_3 - \frac{1}{5}V_4 = 5$
4. $-\frac{1}{5}V_3 + [\frac{1}{5} + \frac{1}{4} + \frac{1}{15}]V_4 - \frac{1}{15}V_5 - \frac{1}{4}V_6 = 0$
5. $-\frac{1}{15}V_4 + [\frac{1}{10} + \frac{1}{15} + \frac{1}{5}]V_5 - \frac{1}{10}V_6 - \frac{1}{5}V_7 = 0$
6. $-\frac{1}{5}V_2 - \frac{1}{4}V_4 - \frac{1}{10}V_5 + [\frac{1}{5} + \frac{1}{4} + \frac{1}{10} + \frac{1}{8}]V_6 - \frac{1}{8}V_7 = 0$
7. $-\frac{1}{4}V_1 - \frac{1}{5}V_5 - \frac{1}{8}V_6 + [\frac{1}{4} + \frac{1}{8} + \frac{1}{5}]V_7 = 0$

What is the diagonal matrix after Gauss–Jordan elimination, and what are the corresponding voltages $V_1 \rightarrow V_7$?

11

Eigenvalues

11.1 ENGINEERS AND EIGENVALUES

Eigenvalues were not invented merely to keep a few mathematicians happy. 'Eigenvalues' form the more advanced phase of the 'matrix' part of the course. You have met eigenvalues before, in principal stresses, or in principal moments of inertia, but perhaps nobody told you.

They have lots of applications in real engineering. Although they represent a facet of matrices, they deserve a separate chapter, in our opinion. If, as we hope, this book is to form a user-friendly environment for practical engineers, then we are faced with a real teaching problem. If we should fail, you would be left with an impression that many engineers unfortunately have, that eigenvalues are really PhD material, i.e. right outside your future scope as an ordinary guy. If we were honest (for a change!) we would be telling you to give eigenvalues a wide berth, to learn just enough to pass your own exams, and then to forget it all with a clear conscience! Those are *not* our sentiments; if we have anti-eigenvalue feelings, yes, the applications are sometimes difficult to grasp, and they tend to be expensive in computing time. That *is* a pity.

11.2 WHAT EVERYBODY SHOULD KNOW

When, under what conditions, can a big matrix be replaced by a scalar? Let's see an example —

$$\begin{bmatrix} 1 & 1 & 2 \\ 1 & 0 & -1 \\ 2 & -1 & 1 \end{bmatrix} \begin{bmatrix} 1 \\ 0 \\ 1 \end{bmatrix} = \begin{bmatrix} 3 \\ 0 \\ 3 \end{bmatrix} = 3 \begin{bmatrix} 1 \\ 0 \\ 1 \end{bmatrix}.$$

The problem is, given a matrix $\mathbf{A}$, how can we find an 'eigenvector' v such that $\mathbf{A}v$ is a scaled-up replica of v, that is $\mathbf{A}v = \lambda v$? So $\lambda = 3$ above, an 'eigenvalue'. λ replaces **A**! But for this vector only. Also

$$\begin{bmatrix} 1 & 1 & 2 \\ 1 & 0 & -1 \\ 2 & -1 & 1 \end{bmatrix} \begin{bmatrix} 1 \\ -1 \\ -1 \end{bmatrix} = -2 \begin{bmatrix} 1 \\ -1 \\ -1 \end{bmatrix}.$$

Yet again:

$$\begin{bmatrix} 1 & 1 & 2 \\ 1 & 0 & -1 \\ 2 & -1 & 1 \end{bmatrix} \begin{bmatrix} 1 \\ 2 \\ -1 \end{bmatrix} = 1 \begin{bmatrix} 1 \\ 2 \\ -1 \end{bmatrix}.$$

There are three cases here, N cases for an $N \times N$ matrix. Of course: for the equations to be consistent, $\mathbf{A}v_i = \lambda_i I v_i$ requires that $\det [\mathbf{A} - \lambda_i I] = 0$:

$$\begin{vmatrix} 1-\lambda & 1 & 2 \\ 1 & 0-\lambda & -1 \\ 2 & -2 & 1-\lambda \end{vmatrix}$$

$$= (1-\lambda) \begin{vmatrix} 0-\lambda & -1 \\ -1 & 1-\lambda \end{vmatrix} - 1 \begin{vmatrix} 1 & -1 \\ 2 & 1-\lambda \end{vmatrix} + 2 \begin{vmatrix} 1 & 0-\lambda \\ 2 & -1 \end{vmatrix}$$

$$= \lambda^3 - 2\lambda^2 - 5\lambda + 6$$

$$= 0 \qquad \text{for } \lambda = 3, -2 \text{ or } 1.$$

You have probably seen this determinant form in some mathematics course. It is of little practical use, except to introduce some of the essential properties of eigenvalues. For example you are unlikely ever to need the eigenvalues of an unsymmetrical matrix, where the λ may be complex even for a real matrix:

$$\begin{vmatrix} 2-\lambda & 1 \\ -1 & 1-\lambda \end{vmatrix} = (2-\lambda)(1-\lambda) - (1)(-1)$$

$$= 0 \text{ for } \lambda = \tfrac{1}{2}(3 \pm i\sqrt{3}).$$

However, the matrix need not be positive definite for the λ_i to be real. (The matrix you have just seen with eigenvalues 3, -2, and 1, was symmetrical but not positive definite.)

The eigenvectors are 'orthogonal' to each other; the scalar products $v_j^T v_j$ are zero, e.g.

$$\begin{bmatrix} 1 & 0 & 1 \end{bmatrix} \begin{bmatrix} 1 \\ -1 \\ -1 \end{bmatrix} = 0.$$

Remember this. To prove it, $\mathbf{A}v_i = \lambda_i v_i$, $\mathbf{A}v_j = \lambda_j v_j$ give the scalar equalities:

$$v_j^T \mathbf{A} v_i = \lambda_i v_j^T v_i$$

$$v_i^T \mathbf{A} v_j = \lambda_j v_i^T v_j.$$

But $v_j^T A v_i$ 'transposes' to give an identical 1×1 matrix, the same scalar quantity, $(v_j^T A v_i)^T = v_i^T A^T v_j$, which by section 10.5 $= v_i^T A v_j$ (because A is symmetric). So therefore our first equation becomes

$$v_i^T A v_j = \lambda_i v_i^T v_j.$$

Subtracting the second equation above gives

$$0 = (\lambda_i - \lambda_j) v_i^T v_j.$$

If $\lambda_i \neq \lambda_j$ the result follows that $v_i^T v_j = 0$. Also $v_i^T A v_j = 0$. Note that if $\lambda_i = \lambda_j$, then

$$A\{k_1 v_i + k_2 v_j\} = \lambda_i \{k_1 v_i + k_2 v_j\}$$

so that any linear combination of v_i and v_j behaves like v_i or v_j alone. Hence, if $\lambda_i = \lambda_j$ and $v_i^T v_j \neq 0$, it is always possible to replace v_j by some linear combination $v_j^* = k_1 v_i + k_2 v_j$ such that $v_j^{*T} v_i = 0$. In this case, we make $v_i^{*T} v_j = 0$ by convention. Eigenvectors are therefore always orthogonal.

11.3 MORE GENERAL KNOWLEDGE

Things you ought to know about eigenvalues—part of 'matrix literacy' which is an important ingredient for success in the computing world of today.

The eigenvectors of A^{-1} are the same as those of A, but the eigenvalues —like the determinant—are inverted. Pre-multiplying by A^{-1} and dividing by λ_i:

$$\lambda_i v_i = A v_i \quad \text{as before}$$
$$A^{-1} v_i = \lambda_i^{-1} v_i.$$

Again, the eigenvalue stands in place of the matrix. This is generally true. You can define other matrix functions, e.g. $\sinh^{-1} A v_i = \sinh^{-1} \lambda_i v_i$—if this amuses you.

Back to work. Let's try and develop some 'indicators' and rules of thumb. We expand 3×3 determinant equation in general:

$$\begin{vmatrix} a - \lambda & b & d \\ b & c - \lambda & e \\ d & e & f - \lambda \end{vmatrix}$$

$$= (a - \lambda) \begin{vmatrix} c - \lambda & e \\ e & f - \lambda \end{vmatrix} - b \begin{vmatrix} b & e \\ d & f - \lambda \end{vmatrix} + d \begin{vmatrix} b & c - \lambda \\ d & e \end{vmatrix}$$

$$= -\lambda^3 + \lambda^2 (a + c + f)$$

$$- \lambda \left\{ \begin{vmatrix} a & b \\ b & c \end{vmatrix} + \begin{vmatrix} a & d \\ d & f \end{vmatrix} + \begin{vmatrix} c & e \\ e & f \end{vmatrix} \right\}$$

$$+ \begin{vmatrix} a & b & d \\ b & c & e \\ d & e & f \end{vmatrix}.$$

We know the three eigenvalues are λ_1, λ_2, and λ_3, so we can equate the above.

$$= -(\lambda - \lambda_1)(\lambda - \lambda_2)(\lambda - \lambda_3)$$
$$= -\lambda^3 + \lambda^2(\lambda_1 + \lambda_2 + \lambda_3) - \lambda(\lambda_1\lambda_2 + \lambda_2\lambda_3 + \lambda_3\lambda_1) + \lambda_1\lambda_2\lambda_3 = 0.$$

So, by equating coefficients:

(i) $\lambda_1 + \lambda_2 + \lambda_3 = a + c + f$ $\qquad\qquad\qquad$ $(3 - 2 + 1 = 1 + 0 + 1)$

(from example at beginning of chapter); i.e.: sum of eigenvalues. = sum of diagonals, 'trace A'! and

(ii) $\lambda_1\lambda_2\lambda_3 = \det [\mathbf{A}]$ $\qquad\qquad\qquad\qquad$ $(-3 \times 2 \times 1 = -6)$;

i.e.: product of eigenvalues = determinant of matrix!

The 'product' rule should remind you of pivots. These properties hold for an $N \times N$ matrix. Another important rule involves the 'signature' of the matrix:

$$\text{Number of} \begin{Bmatrix} \text{positive} \\ \text{zero} \\ \text{negative} \end{Bmatrix} \text{diagonal pivots}$$

$$= \text{Number of} \begin{Bmatrix} \text{positive} \\ \text{zero} \\ \text{negative} \end{Bmatrix} \text{eigenvalues,}$$

in general, provided that you attempt to avoid zero pivots by choosing amongst the remaining diagonals. You should check this rule. A 'positive semidefinite' matrix has some zero pivots/eigenvalues, but none are negative. Because the determinant is zero, you can't invert the matrix. A 'singular' matrix—see Section 10.6.

11.4 TWO-MATRIX EIGENVALUES

A more general form of the problem is $\mathbf{A}v_i = \lambda_i \mathbf{B}v_i$ where the two matrices $\mathbf{A}$ and $\mathbf{B}$ have the same eigenvectors. This equation also reduces to a zero determinant, i.e. an equation for λ_i:

$$\det [\mathbf{A} - \lambda_i \mathbf{B}] = 0.$$

It is just as common in engineering. But following our comments on matrix literacy in section 10.1 we shall not go into the numerical techniques. A simple example:

$$\begin{bmatrix} 1 & 0 & 1 \\ 0 & -1 & -2 \\ 1 & -2 & 3 \end{bmatrix} \begin{bmatrix} 2 \\ 1 \\ 1 \end{bmatrix} = \begin{bmatrix} 3 \\ -3 \\ 3 \end{bmatrix},$$

$$\begin{bmatrix} 1 & -1 & 0 \\ -1 & 2 & -1 \\ 0 & -1 & 2 \end{bmatrix} \begin{bmatrix} 2 \\ 1 \\ 1 \end{bmatrix} = \begin{bmatrix} 1 \\ -1 \\ 1 \end{bmatrix}, \lambda = 3$$

$$\begin{bmatrix} 1 & 0 & 1 \\ 0 & -1 & -2 \\ 1 & -2 & 3 \end{bmatrix} \begin{bmatrix} 1 \\ 2 \\ 1 \end{bmatrix} = \begin{bmatrix} 2 \\ -4 \\ 0 \end{bmatrix},$$

$$\begin{bmatrix} 1 & -1 & 0 \\ -1 & 2 & -1 \\ 0 & -1 & 2 \end{bmatrix} \begin{bmatrix} 1 \\ 2 \\ 1 \end{bmatrix} = \begin{bmatrix} -1 \\ 2 \\ 0 \end{bmatrix}, \lambda = -2$$

$$\begin{bmatrix} 1 & 0 & 1 \\ 0 & -1 & -2 \\ 1 & -2 & 3 \end{bmatrix} \begin{bmatrix} 2 \\ 1 \\ -1 \end{bmatrix} = \begin{bmatrix} 1 \\ 1 \\ -3 \end{bmatrix},$$

$$\begin{bmatrix} 1 & -1 & 0 \\ -1 & 2 & -1 \\ 0 & -1 & 2 \end{bmatrix} \begin{bmatrix} 2 \\ 1 \\ -1 \end{bmatrix} = \begin{bmatrix} 1 \\ 1 \\ -3 \end{bmatrix}, \quad \lambda = 1.$$

Here, λ_i denotes 'A/B' in a particular case, which is as ridiculous as substituting a scalar for a single matrix: nevertheless, it is useful. Check that $v_i^T A v_j = v_i^T B v_j = 0$—a new kind of 'orthogonality'. This problem is really no more difficult. Let's assume $\mathbf{B} = \mathbf{LU}$, the Choleski form, where $\mathbf{L} = \mathbf{U}^T$. Then

$$\mathbf{A} v_i = \lambda_i \mathbf{LU} v_i.$$

Put $\mathbf{U} v_i = w_i$. Then

$$\mathbf{AU}^{-1} w_i = \lambda_i \mathbf{L} w_i$$

$$[\mathbf{L}^{-1} \mathbf{AU}^{-1}] w_i = \lambda_i w_i.$$

The matrix here is symmetric, so the λ_i are real; but this is only possible if we can do Choleski. That is, either **A** or **B** must be 'positive definite' for λ_i real; even 'positive semidefiniteness' will guarantee real roots, and real vectors, provided that values like $\lambda_i = 0/0$ or $1/0$, 'indeterminate', are acceptable. This last comment will make good sense when the physical meaning of these eigenvalues is understood.

11.5 PRELIMINARY MEANINGS

One of the highest forms of art, in numerical methods as the engineer sees them, must be to *recognise* an unfamiliar eigenvalue situation when it arises. This will distinguish the engineer for few mathematicians are likely to do it. We must do it, or it probably won't get done. A nagging problem: suddenly one's instinct bridges the gap, and the complete, elegant solution is revealed. Can this be taught? Probably not, but if we expose you to the idea, it may help you later.

Let's start with an unimportant case, something which probably annoyed you at the time. Remember section 8.3, instability in the thermal conduction problem? We postulated a 'pattern of noise' that would propagate itself; or which, with a slightly longer time step, would slowly increase until the computer overflows in disgust. The rigorous eigenvalue solution would extract all those patterns of 'noise' that would reappear unaltered in the next time step, multiplied by a factor—yes, an eigenvalue! This is the sort of flash of inspiration that we were talking about. If we can calculate the factor, i.e. the eigenvalue, then the criterion as to whether the computer will 'go unstable' with uncontrollable noise, reduces to the question, do *any* of the eigenvalues lie outside the range -1 to 1? Like 'super-roundoff', section 3.2. In detail,

$$[\theta]_{t+\varepsilon} = [\theta]_t + \frac{\varepsilon}{h^2}[\delta^2\theta]_t$$

$$= \left([\mathbf{I}] + \frac{\varepsilon}{h^2}[\mathbf{M}_2]\right)\begin{Bmatrix} \theta_2 \\ \vdots \\ \theta_9 \end{Bmatrix}$$

where

$$\mathbf{M}_2 = \begin{bmatrix} 2 & -1 & 0 & 0 & 0 & 0 & 0 & 0 \\ -1 & 2 & -1 & 0 & 0 & 0 & 0 & 0 \\ 0 & -1 & 2 & -1 & 0 & 0 & 0 & 0 \\ 0 & 0 & -1 & 2 & -1 & 0 & 0 & 0 \\ 0 & 0 & 0 & -1 & 2 & -1 & 0 & 0 \\ 0 & 0 & 0 & 0 & -1 & 2 & -1 & 0 \\ 0 & 0 & 0 & 0 & 0 & -1 & 2 & -1 \\ 0 & 0 & 0 & 0 & 0 & 0 & -1 & 2 \end{bmatrix}.$$

Thus $\mathbf{M}_2$ introduces the second differences, as in section 8.3. The problem in eigenvalue form is to discover the greatest value of the timestep ε, for which the matrix $[\mathbf{I}] + \varepsilon/h^2[\mathbf{M}_2]$ has eigenvalues between -1 and 1. Obviously this problem is of purely academic interest. The pessimistic solution in section 8.3 should be adequate for all practical purposes—and much cheaper!

Our next example is more subtle:

$$\frac{x^2}{\alpha^2} + \frac{y^2}{\beta^2} + \frac{z^2}{\gamma^2} = 1 \tag{11.1}$$

represents an ellipsoid, with its centroid at the origin, and with semi-axes α, β, and γ. We can write the general quadric with the centre at the origin:

$$[x \quad y \quad z]\begin{bmatrix} a & b & d \\ b & c & e \\ d & e & f \end{bmatrix}\begin{bmatrix} x \\ y \\ z \end{bmatrix} = ax^2 + cy^2 + fz^2 + 2bxy + 2dxz + 2eyz = 1$$

—is this an ellipsoid, or a hyperboloid? When one or two of α, β, γ are imaginary we have a hyperboloid—which doesn't have semi-axes in the directions where it goes imaginary!—anyway, in what directions are the axes?

Good guess: yes, the eigenvectors! For convenience, we usually convert these into 'direction cosines' by dividing v_i by $\sqrt{(v_i^T v_i)}$ so that the scaled eigenvectors are of unit length, pointing along the axes. (We can scale v_i as we please: $A(kv_i) = \lambda_i(kv_i)$ for any scalar factor k). Let's write a point on the quadric surface, (xyz), as $k_1 v_1 + k_2 v_2 + k_3 v_3$, where the v are now 'directions cosines'. The equation of the surface becomes

$$[k_1 v_1^T + k_2 v_2^T + k_3 v_3^T][A][k_1 v_1 + k_2 v_2 + k_3 v_3] = 1.$$

We now recall that $v_i^T A v_j = 0$ if $i \neq j$. Further, $v_i^T A v_i = \lambda_i$ because the v are direction cosines and thus $v_i^T v_i = 1$. Hence, the equation reduces to

$$\lambda_1 k_1^2 + \lambda_2 k_2^2 + \lambda_3 k_3^2 = 1. \qquad (11.2)$$

Let's now interpret this geometrically, in terms of scalar products. To reach the point (x, y, z) we go k_1 along v_1, plus k_2 parallel to v_2, then k_3 parallel to v_3, much like the arguments we used in section 4.2. Thus k_1, k_2, and k_3 become the Cartesian coordinates in the axes v_1, v_2, and v_3—which we saw are mutually orthogonal. We have arrived. Comparing equations (11.1) and (11.2), and remembering the relationship between (xyz) and k_1, k_2, and k_3, indicates that the semi-axes are $1/\sqrt{\lambda_i}$ in turn. The problems of principal stresses, principal strains, and principal moments of inertia in dynamics and in bending, and many others, are identical in form. Eigenvalues—you have already used them, without knowing!

But so far we have only seen examples of 'single matrix' eigenvalues. Let's grow up. For a two-matrix example, we could take any vibration problem: the compound pendulum of Fig. 11.1(a) is a simple case. The horizontal deflections being small, CD makes an angle $(w - v)$ to the vertical; with a

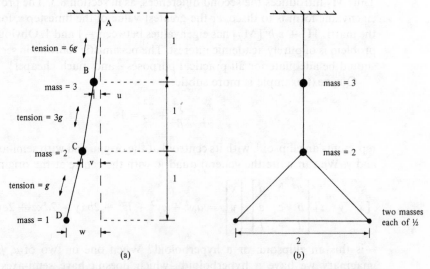

Fig. 11.1 (a) A compound pendulum, three point masses supported by a weightless string. g is acceleration due to gravity. (b) With a strut separating the two masses D, we have a more complicated 'kinetic energy' matrix (section 11.6).

tension of g in CD, this becomes the sideways component. Applying Newton's laws to mass D gives

$$\frac{\ddot{w}}{g} = -w + v.$$

We are attempting to find a natural frequency, ω radians/sec, so we put $u = u_0 \cos \omega t$, $v = v_0 \cos \omega t$, and $w = w_0 \cos \omega t$ so that changing the sign,

$$\frac{\omega^2}{g} w_0 = w_0 - v_0;$$

similarly for mass C

$$2\frac{\ddot{v}}{g} = (w - v) - 3(v - u)$$

giving $2\dfrac{\omega^2}{g} v_0 = -w_0 + 4v_0 - 3u_0$

and for mass B

$$3\frac{\ddot{u}}{g} = 3(v - u) - 6u$$

giving $3\dfrac{\omega^2}{g} u_0 = -3v_0 + 9u_0$.

In matrix notation, a pattern emerges:

$$\frac{\omega^2}{g} \begin{bmatrix} 1 & 0 & 0 \\ 0 & 2 & 0 \\ 0 & 0 & 3 \end{bmatrix} \begin{bmatrix} w_0 \\ v_0 \\ u_0 \end{bmatrix} = \begin{bmatrix} 1 & -1 & 0 \\ -1 & 4 & -3 \\ 0 & -3 & 9 \end{bmatrix} \begin{bmatrix} w_0 \\ v_0 \\ u_0 \end{bmatrix}.$$

The moral of this exercise is that we must be prepared to recognise a two-matrix eigenvalue problem, when it arises. Don't be frightened. The determinant form is

$$\begin{vmatrix} 1 - \dfrac{\omega^2}{g} & -1 & 0 \\ -1 & 4 - \dfrac{2\omega^2}{g} & -3 \\ 0 & -3 & 9 - \dfrac{3\omega^2}{g} \end{vmatrix} = 0$$

$$= 18 - 54\frac{\omega^2}{g} + 36\frac{\omega^4}{g^2} - 6\frac{\omega^6}{g^3}$$

so that $\omega^2/g = 0.46791$, 1.6527, or 3.8794. It would involve more effort to find the modal shapes, i.e. the eigenvectors $[w_0 v_0 u_0]$. You have to solve the three equations $[A - (\omega_i^2/g)B] v_i = 0$. Try it if you like—it's a bore.

11.6 ANOTHER MEANING: QUADRATIC FORMS AND ENERGIES

There is an alternative way of looking at the principal axes of an ellipsoid. The directions can be chosen such that they are perpendicular to the tangent planes, at the points of tangency. In other words, if we move a small distance away from the point of intersection, on the surface of the ellipsoid, then to second order we should be at the same radius from the origin. So we are looking for a point on the surface, whose distance from the origin is stationary. If $(x, y, z) = v$, then $v^T A v = 1$, and we want $\sqrt{(x^2 + y^2 + z^2)} = \sqrt{(v^T v)}$ to be stationary, or being academic, $v^T I v$ where I is the 'identity matrix' (section 10.7).

It is just as easy to argue the stationary conditions for the general two-matrix case, $A v_i = \lambda_i B v_i$, putting $B = I$. We often consider the ratio of the two 'quadratic forms', known as the 'Rayleigh quotient'

$$R = \frac{v^T A v}{v^T B v}.$$

This is a continuous function of the direction v, unless $v^T B v = 0$, and it equals λ_i if v is exactly in the direction v_i. What happens if $v = v_i + dv$? A slight perturbation, v_i is slightly wrong? Interesting:

$$\lambda_i + d\lambda = \frac{(v_i + dv)^T A (v_i + dv)}{(v_i + dv)^T B (v_i + dv)}$$

$$= \frac{v_i^T A v_i + v_i^T A\,dv + dv^T A v_i + dv^T A\,dv}{v_i^T B v_i + v_i^T B\,dv + dv^T B v_i + dv^T B\,dv}$$

$$= \frac{\lambda_i(v_i^T B v_i + v_i^T B\,dv + dv^T B v_i) + dv^T A\,dv}{(v_i^T B v_i + v_i^T B\,dv + dv^T B v_i) + dv^T B\,dv},$$

so that if the terms of dv are of order 10^{-3}, then $d\lambda$ is of order $10^{-6}\lambda_i$, that is, of second order, because $dv^T A\,dv$ and $dv^T B\,dv$ are of order 10^{-6}.

Definition:
[vector]^T [matrix][vector] is known as a 'quadratic form', a scalar function the vector. In the practical cases, involving sizeable matrices, it is usually an energy. Rayleigh's quotient becomes an energy ratio. Many cases emerge in engineering, based on the following elementary concepts.

In mechanics,

kinetic energy $= \frac{1}{2}$ (mass) (velocity)^2

strain energy $= \frac{1}{2}$ (flexibility) (force)^2

or $= \frac{1}{2}$ (stiffness) (deflection)^2

viscous dissipation (power $=$ energy/second)

$= $ (dashpot damping) (velocity)^2

buckling energy (a perturbation) $= -\frac{1}{2}$ (extra stiffness) (deflection)^2.

It follows from the argument above, that if we were calculating a frequency, for example, using Rayleigh's quotient, our answer would be very tolerant of slight errors in v_i, i.e. our assumptions about the deflected shape.

Further, if we are solving the example of Fig. 11.1(b), we can use a short cut in writing the eigenvalue equations. The Lagrange equations in mechanics follow the same philosophy as Rayleigh's quotient recalling strain energy methods in static problems: we think energies rather than equilibrium, i.e. Newton's laws. This is usually much easier. For example, in Fig. 11.1(a) the kinetic energy is

$$\tfrac{1}{2}(3\dot{u}^2 + 2\dot{v}^2 + \dot{w}^2) = \tfrac{1}{2}[\dot{u} \quad \dot{v} \quad \dot{w}] \begin{bmatrix} 3 & 0 & 0 \\ 0 & 2 & 0 \\ 0 & 0 & 1 \end{bmatrix} \begin{bmatrix} \dot{u} \\ \dot{v} \\ \dot{w} \end{bmatrix},$$

and this is the real reason why the matrix appeared in the eigenvalue equation. Similarly, the kinetic energy from Fig. 11.1(b) is

$$\tfrac{1}{2}\{3\dot{u}^2 + 2\dot{v}^2 + \tfrac{1}{2}[2\dot{w}^2] + \tfrac{1}{2}[2(\dot{w} - \dot{v})^2]\}$$

$$= \tfrac{1}{2}[\dot{u} \quad \dot{v} \quad \dot{w}] \begin{bmatrix} 3 & 0 & 0 \\ 0 & 3 & -1 \\ 0 & -1 & 2 \end{bmatrix} \begin{bmatrix} \dot{u} \\ \dot{v} \\ \dot{w} \end{bmatrix},$$

and this more complicated matrix will replace the diagonal matrix in the eigenvalue equation. Energy methods simplify matters in many fields of engineering. It makes everything so much easier, you feel at first that you're cheating.

(b)   In electricity,

ohmic dissipation = (resistance) (current)^2

dielectric energy = $\tfrac{1}{2}$ (capacitance) (voltage)^2

or = $\tfrac{1}{2}$ (charge)^2/(capacitance)

magnetic field energy = $\tfrac{1}{2}$ (inductance) (current)^2.

(c)   In flow through porous media (e.g. groundwater flow d(volume)

d(dissipation) = (permeability) (pressure gradient)^2 d(vol)

(d)   In heat conduction (not directly a dissipation)

d(functional) = (conductivity) (Temperature gradient)^2 d(vol).

In all these cases, with N degrees of freedom the quantities become quadratic forms, usually defining a full matrix. Mostly, these quantities are intrinsically positive, because energies are usually positive, so positive definite matrices are the general rule.

11.7   POSITIVE DEFINITENESS AND RAYLEIGH'S QUOTIENT

Definition:

A matrix is positive definite if the quadratic form is positive for any vector v. This is the important definition in practice. Very often, we deduce that symmetric reduction (or Choleski) is feasible, or that the eigenvalues are positive, simply by considering the physical problem. To argue physically that a matrix is positive definite, is to decide what numerical techniques are feasible! To remind you, we repeat the other definitions:

Definition:
A matrix is positive definite if all the eigenvalues are positive.

Definition:
A matrix is positive definite if all the pivots are positive.
    You should know a variant of this, for hand calculations:

$$\begin{bmatrix} a & b & d \\ b & c & e \\ d & e & f \end{bmatrix} \quad \text{is positive definite, if and only if} \quad a > 0$$

and

$$\begin{vmatrix} a & b \\ b & c \end{vmatrix} > 0$$

and

$$\begin{vmatrix} a & b & d \\ b & c & e \\ d & e & f \end{vmatrix} > 0$$

which is, again, three inequalities. These 'nested' determinants are, in a sense, obvious. Consider this:

$$[0 \quad y \quad z \quad 0] \begin{bmatrix} a & b & d & g \\ b & c & e & h \\ d & e & f & i \\ g & h & i & j \end{bmatrix} \begin{bmatrix} 0 \\ y \\ z \\ 0 \end{bmatrix} = [y \quad z] \begin{bmatrix} c & e \\ e & f \end{bmatrix} \begin{bmatrix} y \\ z \end{bmatrix} > 0$$

implies that $\begin{bmatrix} c & e \\ e & f \end{bmatrix} > 0$, i.e. is positive definite. The other coefficients are not used here. In other words, any diagonal sub-matrix must also be positive definite. It is useful to know this; hence

$$\mathbf{A}_{ii}\mathbf{A}_{jj} > \mathbf{A}_{ij}^2, \quad \text{also} \quad \mathbf{A}_{ii} > 0$$

for any $i \neq j$. Of course, this is not sufficient, but it sometimes helps us to find a mistake.
    For sniffing out and recognising an eigenvalue situation, the following approach occasionally provides the clue. Consider again the Rayleigh quotient, as a 'function' of the vector.

$$R = \frac{v^T \mathbf{A} v}{v^T \mathbf{B} v}.$$

Let's say $[\mathbf{B}] > 0$, positive definite. Then

$$\lambda_{min} \leq R \leq \lambda_{max}.$$

Sometimes we can do this, without actually calculating the eigenvalues, by physical or other arguments. In such cases eigenvalue arguments really are useful. They save you work. You deduce relevant facts without actually calculating anything!

11.8 ZEROS OF QUADRATIC FORMS

Let's return to our quadric surface. If $\mathbf{A}$ is positive definite, we have an ellipsoid. But if $\mathbf{A}$ is 'indefinite' (positive or negative, eigenvalues of mixed sign) we have a hyperboloid. Let's try another reasoning now: what distinguishes a hyperboloid (or a hyperbola in 2D) from an ellipsoid is that it contains

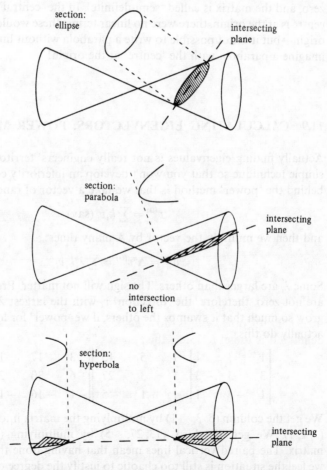

Fig. 11.2 Description of conic sections, which include all three quadratic curves: ellipses, parabolas, and hyperbolas, depending on the angle of the intersecting plane. The sections are drawn as plan views taken perpendicular to the intersecting plane.

'points at infinity'. Consider $v^T A v = 1$, with v very large. Then $v^T A v = 0$ would involve almost the same numbers! Now consider v as of unit length, 'pointing towards' the point at infinity, so that $v^T A v$ is a continuous function of the direction. If $v^T A v$ can be either positive or negative, then we presume there is a zero somewhere between the two regions. Thus an indefinite quadratic form can be zero for real directions. There are points at infinity, in that case. A positive definite or negative definite form cannot be zero. Which proves something, sort of....

We have not mentioned parabolas, which lie between ellipses and hyperbolas: see Fig. 11.2, which shows pictorially how each merges into the other(s). For example, the parabola $x^2 = 1 + y$ becomes

$$[x \quad y] \begin{bmatrix} 1 & 0 \\ 0 & 0 \end{bmatrix} \begin{bmatrix} x \\ y \end{bmatrix} = 1 + y$$

where the matrix is neither positive definite nor indefinite; one eigenvalue is zero, and the matrix is called 'semi-definite'. In the 'central' quadrics, whose centre is at the origin, there were no linear terms; these would have shifted the origin—but it is not possible to write a parabola without linear terms. Try to imagine a parabola with the 'centre' at the origin!

11.9   CALCULATING EIGENVECTORS: POWER METHOD

Actually finding eigenvalues is not really engineers' territory. We include a simple technique so that you won't develop an inferiority complex. The idea behind the 'power' method is that we take a vector of random numbers:

$$v^{(0)} = \sum k_i v_i \text{ (say)}$$

and then we multiply the vector by **A** many times:

$$v^{(n)} = \mathbf{A}^n v^{(0)} = \sum \lambda_i^n k_i v_i.$$

Some λ_i are larger than others. The sign will not matter. Provided that the k_i are not zero, therefore, the 'direction' v_i with the largest λ_i, will eventually grow so much that it swamps the others, if we 'power' for long enough. Let's actually do this.

$$\begin{bmatrix} 1 & 3 & 1 \\ 3 & 3 & -3 \\ 1 & -3 & 1 \end{bmatrix} \begin{matrix} 1 \\ 1 \\ 1 \end{matrix} \begin{matrix} 5 \\ 3 \\ -1 \end{matrix} \begin{matrix} 13 \\ 27 \\ -5 \end{matrix} \left\| \begin{matrix} 3 \\ 6 \\ -1 \end{matrix} \right. \begin{matrix} 17 \\ 30 \\ -16 \end{matrix} \begin{matrix} 1 \\ 2 \\ -1 \end{matrix} \left\| \begin{matrix} 6 \\ 12. \\ -6 \end{matrix} \right.$$

We get the column $(5, 3, -1)$ by multiplying the matrix into the first column, $(1, 1, 1)$. We get the third, $(13, 27, -5)$ by multiplying $(5, 3, -1)$ by the matrix. The paired vertical lines mean that having done the multiplication, we feel the situation is still too chaotic to justify the degree of accuracy, so we 'round-off' by approximately scaling down the vector, trying to keep roughly the same ratio between the numbers. Here we divided by 4.5 and rounded. A scaled-down vector tells the same story. We repeat the exercise until we discover the 'eigenvalue situation':

$$\begin{bmatrix} 1 & 3 & 1 \\ 3 & 3 & -3 \\ 1 & -3 & 1 \end{bmatrix} \begin{bmatrix} 1 \\ 2 \\ -1 \end{bmatrix} = \begin{bmatrix} 6 \\ 12 \\ -6 \end{bmatrix} = 6 \begin{bmatrix} 1 \\ 2 \\ -1 \end{bmatrix}, \lambda_1 = 6.$$

This happens with simple numbers, because we are kind-hearted teachers! Don't expect nature to be as kind.

Next, we 'zoo' the matrix: the physical explanation is that we create a matrix whose response to v_2 and v_3 is as before but the eigenvalue associated with v_1 is shifted to zero. With v_1 out of the competition, 'powering' again will lead to v_2. First the maths, then an example.

$$\mathbf{A}^* = \mathbf{A} - \lambda_1 \frac{v_1 v_1^T}{(v_1^T v_1)}.$$

Thus, for example

$$\mathbf{A}^* v_2 = \mathbf{A} v_2 - \lambda_1 \frac{v_1 v_1^T}{v_1^T v_1} v_2$$

$$= \lambda_2 v_2 - \lambda_1 \frac{v_1 (v_1^T v_2)}{v_1^T v_1}$$

$$= \lambda_2 v_2 \quad (\text{because } v_1^T v_2 = 0),$$

but

$$\mathbf{A}^* v_1 = \mathbf{A} v_1 - \frac{\lambda_1 [v_1 v_1^T]}{v_1^T v_1}$$

$$= \lambda_1 v_1 - \frac{\lambda_1 v_1 (v_1^T v_1)}{v_1^T v_1}.$$

$$= 0$$

Thus, 'zooing' creates the matrix $\mathbf{A}^*$. The method is sometimes called Hotelling's deflation. As we have shown, $\mathbf{A}^*$ does not react to v_1 but will lead us to v_2. Let's actually create $\mathbf{A}^*$ for our example:

$$\lambda_1 = 6, v_1 = (1, 2, -1)$$

and

$$v_1^T v_1 = 1^2 + 2^2 + 1^2$$

$$\mathbf{A}^* = \begin{bmatrix} 1 & 3 & 1 \\ 3 & 3 & -3 \\ 1 & -3 & 1 \end{bmatrix} - \frac{6}{1^2 + 2^2 + 1^2} \begin{bmatrix} 1 \\ 2 \\ -1 \end{bmatrix} \begin{bmatrix} 1 & 2 & -1 \end{bmatrix}$$

$$= \begin{bmatrix} 1 & 3 & 1 \\ 3 & 3 & -3 \\ 1 & -3 & 1 \end{bmatrix} - \begin{bmatrix} 1 & 2 & -1 \\ 2 & 4 & -2 \\ -1 & -2 & 1 \end{bmatrix} = \begin{bmatrix} 0 & 1 & 2 \\ 1 & -1 & -1 \\ 2 & -1 & 0 \end{bmatrix}.$$

Now we 'power' again, using $\mathbf{A}^*$ as the 'starting' matrix. It is often better to start with the penultimate vector in the previous powering:

$$
\begin{bmatrix} 0 & 1 & 2 \\ 1 & -1 & -1 \\ 2 & -1 & 0 \end{bmatrix}
\begin{array}{cc} 17 & -2 \\ 30 & 3 \\ -16 & 4 \end{array}
\begin{Vmatrix} -1 & 5 & -10 \\ 1 & -4 & 12 \\ 2 & -3 & 14 \end{Vmatrix}
\begin{array}{cc} -1 & 3 \\ 1 & -3. \\ 1 & -3 \end{array}
$$

So now we have the eigenvalue situation again: check

$$
\begin{bmatrix} 1 & 3 & 1 \\ 3 & 3 & -3 \\ 1 & -3 & 1 \end{bmatrix}
\begin{bmatrix} -1 \\ 1 \\ 1 \end{bmatrix} =
\begin{bmatrix} 3 \\ -3 \\ -3 \end{bmatrix}, \lambda_2 = -3.
$$

So now we have $\lambda_2 = -3$, $v_2 = (-1, 1, 1)$ and $v_2^\mathsf{T} v_2 = 1^2 + 1^2 + 1^2$.

Let's 'zoo' it again. $\mathbf{A}^{**}$ comes from $\mathbf{A}^*$ using v_2 and λ_2 just as $\mathbf{A}^*$ came from $\mathbf{A}$ using v_1 and λ_1:

$$
\mathbf{A}^{**} = \begin{bmatrix} 0 & 1 & 2 \\ 1 & -1 & -1 \\ 2 & -1 & 0 \end{bmatrix} - \frac{-3}{1^2 + 1^2 + 1^2} \begin{bmatrix} -1 \\ 1 \\ 1 \end{bmatrix} [-1 \quad 1 \quad 1]
$$

$$
= \begin{bmatrix} 0 & 1 & 2 \\ 1 & -1 & -1 \\ 2 & -1 & 0 \end{bmatrix} - \begin{bmatrix} -1 & 1 & 1 \\ 1 & -1 & -1 \\ 1 & -1 & -1 \end{bmatrix} = \begin{bmatrix} 1 & 0 & 1 \\ 0 & 0 & 0 \\ 1 & 0 & 1 \end{bmatrix}.
$$

By doing this we remove the influence of v_2. Powering again will reveal v_3:

$$
\begin{bmatrix} 1 & 0 & 1 \\ 0 & 0 & 0 \\ 1 & 0 & 1 \end{bmatrix}
\begin{array}{cc} -10 & 4 \\ 12 & 0 \\ 14 & 4 \end{array}
\begin{Vmatrix} 1 & 2 \\ 0 & 0. \\ 1 & 2 \end{Vmatrix}
$$

Check:

$$
\begin{bmatrix} 1 & 3 & 1 \\ 3 & 3 & -3 \\ 1 & -3 & 1 \end{bmatrix}
\begin{bmatrix} 1 \\ 0 \\ 1 \end{bmatrix} =
\begin{bmatrix} 2 \\ 0 \\ 2 \end{bmatrix}, \quad \lambda_3 = 2.
$$

'Zoo' for the last time:

$$
\mathbf{A}^{***} = \begin{bmatrix} 1 & 0 & 1 \\ 0 & 0 & 0 \\ 1 & 0 & 1 \end{bmatrix} - \frac{2}{1^2 + 0^2 + 1^2} \begin{bmatrix} 1 \\ 0 \\ 1 \end{bmatrix} [1 \quad 0 \quad 1] = \begin{bmatrix} 1 & 0 & 1 \\ 0 & 0 & 0 \\ 1 & 0 & 1 \end{bmatrix}
$$

$$
= \begin{bmatrix} 0 & 0 & 0 \\ 0 & 0 & 0 \\ 0 & 0 & 0 \end{bmatrix}
$$

This is fun. But it's useful too: working backwards

$$\mathbf{A} = \frac{2}{1^2 + 0^2 + 1^2} \begin{bmatrix} 1 \\ 0 \\ 1 \end{bmatrix} \begin{bmatrix} 1 & 0 & 1 \end{bmatrix} + \frac{-3}{1^2 + 1^2 + 1^2} \begin{bmatrix} -1 \\ 1 \\ 1 \end{bmatrix} \begin{bmatrix} -1 & 1 & 1 \end{bmatrix}$$

$$+ \frac{6}{1^2 + 2^2 + 1^2} \begin{bmatrix} 1 \\ 2 \\ -1 \end{bmatrix} \begin{bmatrix} 1 & 2 & -1 \end{bmatrix}$$

$$= \sum \lambda_i \frac{v_i v_i^{\mathrm{T}}}{v_i^{\mathrm{T}} v_i}$$

So we have 'discovered' (the hard way) that matrices are composed of their eigenvalues and eigenvectors by a subtle and thought-provoking formula. This is how we constructed our examples for this course!

'Black boxes' (or our program EIG) will probably give you all the eigenvectors you will ever need. There is little point for an engineer to learn lots of sophisticated techniques. The 'powering' method has nasty snags, but it is realistic, in that it works for big matrices. It used to be the standard method, and some of its clever offspring are still widely used.

But it is a good idea to know in advance what can go wrong, in eigenvalue calculations.

(a) If two eigenvalues are nearly equal, you have a roundoff problem. Sophisticated theories do not help. Probably the ill-conditioning is a physical phenomenon. That is, if you change the geometry, or the masses of the parts, etc., very slightly, then the eigenvectors will change far more than anybody would expect. That is one reason why we said that practical engineers should know something about the behaviour of eigenvalues. The two eigenvectors are ill-defined, and roundoff causes a lot of trouble.

(b) If an eigenvalue is very small, then its eigenvector is ill-defined. But in this case you can probably find it accurately, by making it orthogonal to the others.

So much for the glitches. You *must* know what can go wrong. Meanwhile, accentuate the positive! An engineer should recognise eigenvalue country, and quickly feel at home in unexplored territory. This may not come as easily, but with the opportunities to calculate what was not possible in the pre-electronic age, you should be in a position to take up any new opportunities that present themselves, and to gain experience. Otherwise, what is the point in an engineer learning about numerical methods? Eigenvalues, in particular.

EXERCISES

11.1. Find the eigenvalues and eigenvectors of the matrices below by powering and zooing. Apply as many checks as you can.

$$
\text{(i)} \quad
\begin{bmatrix}
0 & 8 & -2 \\
8 & 10 & -8 \\
-2 & -8 & 0
\end{bmatrix}
\qquad
\text{(ii)} \quad
\begin{bmatrix}
5 & -1 & -1 \\
-1 & 1 & 3 \\
-1 & 3 & 1
\end{bmatrix}
$$

$$
\text{(iii)} \quad
\begin{bmatrix}
2 & -1 & -1 \\
-1 & 4 & 1 \\
-1 & 1 & 2
\end{bmatrix}
\qquad
\text{(iv)} \quad
\begin{bmatrix}
1 & -3 & 1 \\
-3 & -3 & 3 \\
1 & 3 & 1
\end{bmatrix}.
$$

11.2. (i) For many mechanical systems, the equations of free vibration can be written in the form $[m]\{\ddot{x}\} - [k]\{x\} = 0$. For the pendulum shown, the Lagrange equation yields

$$
[m] = \frac{mgl}{2}\begin{bmatrix} 6 & 0 \\ 0 & 1 \end{bmatrix}
$$

$$
[k] = \frac{ml^2}{6}\begin{bmatrix} 54 & 9 \\ 9 & 2 \end{bmatrix}.
$$

Determine the natural frequencies of vibration and from experience (experiment?!!) state which is associated with which mode.

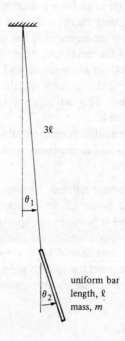

3ℓ

θ_1

θ_2 uniform bar
length, ℓ
mass, m

(ii) For the system shown on the next page, determine the two matrices [m] and [k] and the natural frequencies of vibration.

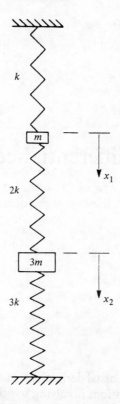

11.3. A physical problem gives rise to the following eigenvalue equation:

$$\begin{bmatrix} 3 & 5 \\ 5 & 8 \end{bmatrix} \begin{bmatrix} x \\ y \end{bmatrix} = \lambda \begin{bmatrix} 3 & 3 \\ 3 & 2 \end{bmatrix} \begin{bmatrix} x \\ y \end{bmatrix}.$$

Are there any reasons for doubt?

11.4. The stress tensor at a particular point in a structural element, relative to an orthogonal axis set (x, y, z), is:

$$\begin{bmatrix} -2.67 & 0.00 & -1.60 \\ 0.00 & -3.11 & -1.08 \\ -1.60 & -1.08 & -11.19 \end{bmatrix}.$$

Determine eigenvalues and eigenvectors. What do they represent?

12

Partial differential equations

12.1 SCOPE

We discussed 'partial derivatives' in section 2.2. Partial differential equations (PDE) are equations involving the sign ∂, instead of d, as in Chapter 8. It is strange to be introducing so important an idea so late, or is it?

Actually, we have seen partial differential equations already: transient heat conduction along a bar, and over a plate (sections 8.2 and 8.4). The physical details of everything that happens around us are governed by PDEs. You might be sitting on a beach, watching the infinite variety of the waves breaking, as if the tops want to move faster than the water underneath; then, when a wave has broken, it washes back into the sea, with a multitude of tiny steplike waves. You can't help getting poetic. But it may cure any human arrogance you may tend to develop, to ponder thus: events of such complexity will *never* be modelled adequately by our computers. Again, mathematicians have shown that the number of different patterns of turbulence which can develop in a wind-tunnel are of order (Reynold's number)$^{2.5}$. Reynold's numbers of 10^5 are common! Nevertheless, there is much talk about using supercomputers instead of wind-tunnels to design aircraft, etc. Presumably those concerned already have techniques for simplifying the problem sufficiently. No doubt the computer will help determine which are the important experiments to perform. When experimental aerodynamics becomes obsolete the computer revolution in engineering will remain incomplete. The problems of designing fusion reactors will be much more difficult again....

PDEs are horribly important, and sometimes they are horribly difficult. Some mathematicians spend their lives investigating the existence of solutions to PDEs, or their uniqueness, or the probable errors of numerical solutions, or even elaborating complicated analytic solutions. Most engineers regard such activities as pointless, academic, or just plain silly. But that's not quite fair—irrelevant, not silly. What the engineer wants to see is the answers, not hear a philosophical discussion.

However, don't presume that engineers are never expected to think! Mathematicians are often surprisingly inept, when it comes to understanding physically what they are doing, despite their high I.Q!! So inevitably, engineers are expected to fill the gap. Let's start the training.

12.2   LONGITUDINAL WAVES IN A BAR

Let's try to gain some experience at developing the PDE that describes a simple and clearly defined physical phenomenon—not solving it, for the moment. It's the same sort of problem that we met in section 8.2, but we approach it in a more mature, adult fashion by considering the slice x to $x + dx$ in Fig. 12.1. Assume again that the bar has unit cross-section. If at x, the deflection is u, then at $x + dx$ the deflection will be $u + (\partial u/\partial x)\,dx$—Taylor's series, two terms. Therefore the tension is E times (strain) and the strain is the change in length of the region x to $x + dx$ divided by the original length, dx:, that is, the strain is $\partial u/\partial x$. Therefore the tension in the bar at position x is $E\,\partial u/\partial x$.

Now look at it in a slightly different way: dx was small, so $E\,\partial u/\partial x$ is the tension at x, and is itself a function of x. At $x + dx$ the tension is $E\{\partial u/\partial x + (\partial^2 u/\partial x^2)\,dx\}$—Taylor again! The elementary region x to $x + dx$ is pulled to the left by the force $E\,\partial u/\partial x$, and so the right by the force $E\{\partial u/\partial x + (\partial^2 u/\partial x^2)\,dx\}$. The net force is $E(\partial^2 u/\partial x^2)\,dx$, to the right. The mass is $\rho\,dx$, because we have unit cross-section. Therefore we can apply Newton's laws:

$$F = ma.$$

The acceleration of course is $\partial^2 u/\partial t^2$, so

$$E\,\frac{\partial^2 u}{\partial x^2}\,dx = \rho\,\frac{\partial^2 u}{\partial t^2}\,dx$$

and cancelling dx,

$$E\,\frac{\partial^2 u}{\partial x^2} = \rho\,\frac{\partial^2 u}{\partial t^2}. \tag{12.1}$$

This is the equation that describes wave propagation along a slender bar. End of exercise. Perhaps we shall solve this PDE later! It is an exceptional case.

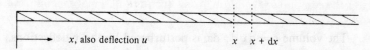

x, also deflection u x $x + dx$

Fig. 12.1   We take a thin rod, and we tap it horizontally, thus exciting a wave which travels along it. There are no vertical movements. Deflections u, in the x-direction, stretch/compress the bar and cause deflections u elsewhere, a moment later.

12.3 SOUND WAVES IN AIR

We choose the next example, not only because it forms a good extension of (12.1), but also—and mainly—because it introduces you to the remarkable power of ∇, the vector differential operator:

$$\nabla = \hat{x}\,\frac{\partial}{\partial x} + \hat{y}\,\frac{\partial}{\partial y} + \hat{z}\,\frac{\partial}{\partial z}$$

where $\hat{x}$, $\hat{y}$, and $\hat{z}$ are the unit vectors in the x, y, and z directions.

Sound waves: we assume they are of small amplitude—so that we have constant ambient air pressure, p, perturbated by small pressure differences, q. Thus q resembles u, the small changes in position along the rod; but now we are in three dimensions, and we must speak of the 'displacement' of the molecules, the vector $\vec{\delta}$. Consider the motion of an element of air $dx\,dy\,dz$. It is the same as the element of rod, in that the effect of the pressure $p + q$ on the faces $dy\,dz$ is to give an x-component of the force $(\partial q/\partial x)\,dx\,dy\,dz$ (note, p contributes no accelerating force). Thus, since we are speaking of positive pressure, Newton gives

$$\ddot{\vec{\delta}}\rho\,dx\,dy\,dz = -\left(\hat{x}\,\frac{\partial q}{\partial x} + \hat{y}\,\frac{\partial q}{\partial y} + \hat{z}\,\frac{\partial q}{\partial z}\right)dx\,dy\,dz$$

or, cancelling $dx\,dy\,dz$

$$\rho\ddot{\vec{\delta}} = -\nabla q. \tag{12.2}$$

In order to develop a PDE in one variable, we must now express q in terms of $\vec{\delta}$. Note that q is a small change in p, effectively dp, so with volume v under adiabatic compression

$$(p + q)(v + dv)^{\gamma} = \text{constant (normally } \gamma = 1.4)$$

(—remember your adiabatic gas law, $pv^{\gamma} = \text{constant}$?).

Or writing the total differential involving q and dv,

$$qv^{\gamma} + p\gamma v^{\gamma - 1}\,dv = 0$$

(derive this from the first equation using the Binomial series or take the differential of the gas law).

So

$$q = -\gamma p\,\frac{dv}{v}.$$

The volume $v = dx\,dy\,dz$, is perturbated by the x-deflections.

$$dv = dx\,dy\,dz\,\frac{\partial \delta_x}{\partial x}$$

—much like du/dx in the bar. Including also the compression in the y and z directions, the total dv is

$$\left(\frac{\partial \delta_x}{\partial x} + \frac{\partial \delta_y}{\partial y} + \frac{\partial \delta_z}{\partial z}\right) dx\, dy\, dz$$

$$= \left(\hat{x}\frac{\partial}{\partial x} + \hat{y}\frac{\partial}{\partial y} + \hat{z}\frac{\partial}{\partial z}\right) \cdot (\hat{x}\delta_x + \hat{y}\delta_y + \hat{z}\delta_z)\, dx\, dy\, dz$$

$$= \nabla \cdot \vec{\delta}\, dx\, dy\, dz$$

$$= v\nabla \cdot \vec{\delta}$$

using the scalar product, in a cunning way that you may not have seen. Thus from before,

$$q = -\gamma p \nabla \cdot \vec{\delta}. \tag{12.3}$$

Differentiating (12.3) twice with time:

$$-\nabla \cdot \ddot{\vec{\delta}} = \frac{\ddot{q}}{\gamma p}. \tag{12.4}$$

Changing the sign of (12.2), pre-multiplying by '∇', and substituting from (12.4), we can eliminate $\vec{\delta}$:

$$-\rho\nabla \cdot \ddot{\vec{\delta}} = \nabla^2 q = \frac{\rho\ddot{q}}{\gamma p}.$$

Job done. Isn't it amazing, how simple it looks!

12.4   WAVE EQUATIONS IN GENERAL

Now we must think about it. How are waves transmitted? What is a wave? That at least seems easy enough. A wave is a phenomenon that transmits information, i.e. fine detail, a signal. A sharp pulse remains a sharp pulse. The criterion is what happens to the fine detail. In thermal conduction along a bar, the sharp discontinuity at the ends is lost immediately, and all trace of it is lost eventually as the curve gets smoother.

Surprisingly, we can write the general solution of (12.1): let $u = f(x - ct) + g(x + ct)$, where f and g are any functions whatsoever. Then $\partial^2 u/\partial x^2 = f''(x - ct) + g''(x + ct)$ where the primes mean that we differentiate with respect to whatever is in the bracket. But to find $\partial^2 f/\partial t^2$, for example, we must use the chain rule twice:

$$\frac{\partial f}{\partial t} = \frac{\partial f}{\partial(x - ct)} \cdot \frac{\partial(x - ct)}{\partial t} = -cf'(x - ct), \qquad \frac{\partial^2 f}{\partial t^2} = c^2 f''(x - ct)$$

so that

$$\frac{\partial^2 u}{\partial t^2} = c^2 u''.$$

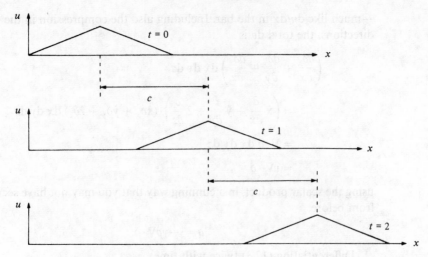

Fig. 12.2   The progress of a wave along a bar. The pattern of compressions does not change as the wave progresses; the sharp features remain, according to theory. There is no distortion of the 'signal'.

But $\rho\ddot{u} = E\,\partial^2 u/\partial x^2 = Eu''$, so that $c = \sqrt{E/\rho}$, the speed of sound. Fig. 12.2 illustrates what happens if the function g is absent, and if f has characteristic features; in theory, they are transmitted unchanged with time—there is no deterioration in the signal. We have demonstrated true wave behaviour, using what is really very elementary calculus. It is probably easier to visualise Fig. 12.2 as the response of a stretched string to a sudden upward 'jerk' at the end; in that case $c = \sqrt{(\text{tension})/(\text{mass per unit length})}$, and the PDE is essentially the same.

12.5 HYPERBOLIC, PARABOLIC AND ELLIPTIC EQUATIONS

This brings us to the central issue: how can we recognise mathematically, from the PDE itself, which class of phenomenon we are dealing with? Let's consider the general equation:

$$a\,\frac{\partial^2 u}{\partial x^2} + c\,\frac{\partial^2 u}{\partial y^2} + f\,\frac{\partial^2 u}{\partial z^2} + 2b\,\frac{\partial^2 u}{\partial x\,\partial y} + 2e\,\frac{\partial^2 u}{\partial y\,\partial z} + 2d\,\frac{\partial^2 u}{\partial x\,\partial z} = \text{etc.}$$

It is better to write this in matrix form:

$$\begin{bmatrix} \dfrac{\partial}{\partial x} & \dfrac{\partial}{\partial y} & \dfrac{\partial}{\partial z} \end{bmatrix} \begin{bmatrix} a & b & d \\ b & c & e \\ d & e & f \end{bmatrix} \begin{bmatrix} \partial/\partial x \\ \partial/\partial y \\ \partial/\partial z \end{bmatrix} u = \text{etc.}$$

From the point of view of classification, any first derivatives need not concern us. As with the ellipsoids, section 11.5, we now transform the axes. Consider the total differential of any function w,

$$dw = \frac{\partial w}{\partial x}\,dx + \frac{\partial w}{\partial y}\,dy + \frac{\partial w}{\partial z}\,dz$$

$$= [dx \; dy \; dz] \begin{bmatrix} \partial/\partial x \\ \partial/\partial y \\ \partial/\partial z \end{bmatrix} w.$$

As before, let the position $(x\,y\,z) = k_1 v_1 + k_2 v_2 + k_3 v_3$ where v_1, v_2, and v_3 are the eigenvectors with $x\,y\,z$ components v_{1x} etc. Thus

$$dw = [dk_1 \; dk_2 \; dk_3] \begin{bmatrix} v_{1x} & v_{1y} & v_{1z} \\ v_{2x} & v_{2y} & v_{2z} \\ v_{3x} & v_{3y} & v_{3z} \end{bmatrix} \begin{bmatrix} \partial/\partial x \\ \partial/\partial y \\ \partial/\partial z \end{bmatrix} w$$

$$= [dk_1 \; dk_2 \; dk_3] \begin{bmatrix} \partial/\partial k_1 \\ \partial/\partial k_2 \\ \partial/\partial k_3 \end{bmatrix} w.$$

This is true for any $[dk_1 \; dk_2 \; dk_3]$, so

$$\begin{bmatrix} v_{1x} & v_{1y} & v_{1z} \\ v_{2x} & v_{2y} & v_{2z} \\ v_{3x} & v_{3y} & v_{3z} \end{bmatrix} \begin{bmatrix} \partial/\partial x \\ \partial/\partial y \\ \partial/\partial z \end{bmatrix} w = \begin{bmatrix} \partial/\partial k_1 \\ \partial/\partial k_2 \\ \partial/\partial k_3 \end{bmatrix} w$$

$$\begin{bmatrix} \partial/\partial x \\ \partial/\partial y \\ \partial/\partial z \end{bmatrix} w = \begin{bmatrix} v_{1x} & v_{1y} & v_{1z} \\ v_{2x} & v_{2y} & v_{2z} \\ v_{3x} & v_{3y} & v_{3z} \end{bmatrix}^{-1} \begin{bmatrix} \partial/\partial k_1 \\ \partial/\partial k_2 \\ \partial/\partial k_3 \end{bmatrix} w = \begin{bmatrix} v_{1x} & v_{2x} & v_{3x} \\ v_{1y} & v_{2y} & v_{3y} \\ v_{1z} & v_{2z} & v_{3z} \end{bmatrix} \begin{bmatrix} \partial/\partial k_1 \\ \partial/\partial k_2 \\ \partial/\partial k_3 \end{bmatrix} w.$$

The matrix inversion is achieved simply by transposing, as here, for any 'orthogonal' matrix; see below.

Definition:
Formally, a matrix $[\theta]$ is 'orthogonal' if $[\theta]\,v = w$ gives $w^{\mathrm{T}} w = v^{\mathrm{T}} v$. That is, 'distances' are unchanged. Typically $[\theta]$ represents a rotation, for example:

$$[\theta] = \begin{bmatrix} \cos\theta & \sin\theta & 0 \\ -\sin\theta & \cos\theta & 0 \\ 0 & 0 & 1 \end{bmatrix}$$

represents a rotation of the x, y axes through angle θ anti-clockwise about the z axis, i.e.: in the x–y plane with z unchanged:

$$\begin{bmatrix} x^1 \\ y^1 \\ z^1 \end{bmatrix} = [\theta] \begin{bmatrix} x \\ y \\ z \end{bmatrix} \qquad (w = [\theta]v).$$

Check that the distance from the origin to some point $P(x, y, z)$ remains unchanged.

Returning to our definition, $w^T w = v^T \theta^T \theta v$, which from above $= v^T I v$, so that $\theta^T \theta = I$, hence $\theta^T = \theta^{-1}$.

Applying this to

$$[\partial/\partial x \quad \partial/\partial y \quad \partial/\partial z][A] \begin{bmatrix} \partial/\partial x \\ \partial/\partial y \\ \partial/\partial z \end{bmatrix} u = \text{etc.}$$

gives

$$[\partial/\partial k_1 \quad \partial/\partial k_2 \quad \partial/\partial k_3] \begin{bmatrix} v_1 & v_1 & v_1 \\ v_2 & v_2 & v_2 \\ v_3 & v_3 & v_3 \end{bmatrix} A \begin{bmatrix} v_1 & v_2 & v_3 \\ v_1 & v_2 & v_3 \\ v_1 & v_2 & v_3 \end{bmatrix} \begin{bmatrix} \partial/\partial k_1 \\ \partial/\partial k_2 \\ \partial/\partial k_3 \end{bmatrix} u.$$

The triple matrix product reduces to a diagonal matrix, as with the quadric surfaces. If v_1, v_2, and v_3 have been normalised, so that $v_i^T v_i = 1$, the PDE reduces to

$$\lambda_1 \frac{\partial^2 u}{\partial k_1^2} + \lambda_2 \frac{\partial^2 u}{\partial k_2^2} + \lambda_3 \frac{\partial^2 u}{\partial k_3^2} = \text{etc.}$$

If all of λ_1, λ_2, and λ_3 are positive, we have an 'elliptic' equation: if one or two of them are negative, we have a 'hyperbolic' equation. See sections 11.5 and 11.8. Positive definiteness again. Powerful things, these eigenvalues!

12.6   GENERAL DISCUSSION

So we have achieved our main objective. For this fairly general class of PDE we can distinguish wave equations as the 'hyperbolic' subclass. Perhaps it will help you to remember the central idea, if you recall that hyperboloids contain points (x, y, z) at infinity, along the asymptotes. Likewise, hyperbolic equations can convey sharp points, as in Fig. 12.2, where $\partial^2 u/\partial t^2$ and $\partial^2 u/\partial x^2$ are infinite—singularities which are not imposed directly by the boundary conditions, and which propagate themselves indefinitely. 'Infinite points' in hyperboloids become 'infinitely sharp details' in the solutions to hyperbolic PDE. Think about 'shock waves' in supersonic aircraft, etc.

This contrasts with the life history of the initial singularity in the thermal conduction problem, in section 8.2, which disappeared very quickly; that was a 'parabolic' equation. It is the fact that singularities are not necessarily 'local' phenomena that makes hyperbolic equations so treacherous and difficult to solve numerically. For example, a stressing problem on a thin membrane is sometimes governed by a hyperbolic equation. Such solutions are academic; but a thin shell may resemble a membrane, and on that account may cause unexpected trouble.

On the other hand, hyperbolic equations enable us to gather signals from very distant quasars, and to communicate between continents and with spacecraft as they leave the solar system. Some men can even recognise an attractive girl at a great distance. Hyperbolic equations make life much more interesting. Also it is difficult to imagine powerful computers without phenomena governed by hyperbolic equations.

To emphasise the point, we quote from a standard text of 20 years ago†, on numerical methods:

The above examples manifest a feature which distinguishes the formulation of problems involving hyperbolic and parabolic equations from those involving elliptic equations, namely that the boundary conditions are commonly specified on an open boundary. For elliptic equations it is usual to have closed boundaries. A typical problem in the elliptic field is: 'Given u on a closed curve, to find u inside that curve to satisfy the boundary condition and the differential equation

$$\frac{\partial^2 u}{\partial x^2} + \frac{\partial^2 u}{\partial y^2} = 0$$

inside the curve'. An interesting feature of this problem is that even if the values of u on the boundary are discontinuous at a number of points, the solution u has derivatives of all orders in both x and y inside the curve. This is in striking contrast to the hyperbolic case, where boundary conditions with discontinuities in the derivatives give rise to solutions with discontinuities in the derivatives. The well-behaved nature of the solution in the elliptic case has the result that finite-difference techniques are far less likely to lead to difficulties, and it is usually quite safe to use a rectangular mesh of points. For hyperbolic equations the possibility of having discontinuities in the second derivative across the characteristics makes the use of a rectangular mesh rather hazardous...

At the time when the above was written, it was usual to refer to parabolic and hyperbolic equations as 'marching' (in time) and to elliptic equations as 'jurying', meaning presumably that what happens at a point is a weighted mean of what would happen if the various influences acted on a simplified model. In other words, don't expect surprises.

Of course, this leaves untouched the general question of classifying PDEs which may be more complicated; for example, nonlinear equations, which can give weird phenomena like 'solitons' (these are solitary pulses, which spread indefinitely with time). A familar case of solitons is the hydraulic jump, a step-like wave which progresses down a river (e.g. the 'Severn bore') or a canal. You see this while you are washing the dishes, very frequently. Again, we have the equation for bending waves, in a thin plate:

$$\frac{Et^3}{12(1-v^2)} \nabla^4 w = \rho t \ddot{w}$$

which we would not expect you to develop! The crazy thing here is that the wave velocity c depends on ω! The same is true of light, except in a vacuum—this plays havoc with the 'signal'. All sorts of complications.

We drop the whole subject of hyperbolic PDE, as being too difficult to compute in real engineering for the moment. One day, perhaps, engineers will have programs to model hyperbolic phenomena realistically; if and when this happens, it will be useful for you to be able to recognise simple wave equations. So let's now concentrate on the relatively easy ones, elliptic PDE.

---

† *Modern Computing Methods*, National Physical Laboratory, Notes on Applied Science No. 16; Her Majesty's Stationery Office, London, 1961 (quote from page 111).

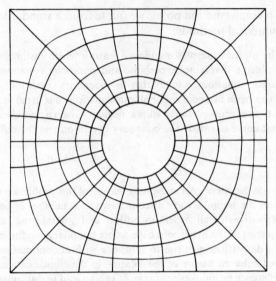

Fig. 12.3   Equipotentials and flow lines in a solution to $\nabla^2 u = 0$. Because there is no temperature gradient along any equipotential, and therefore no flow, the flow lines are strictly orthogonal to the equipotentials.

12.7   TUBES OF FLOW

Some 40 years ago, an Englishman named Southwell organised a team of 'computers', who diligently solved quite complicated elliptic problems for him, using a technique that he called 'relaxation'. Long before that, however, people were solving e.g. static thermal conduction problems fairly successfully, by a technique based on flow lines (Fig. 12.3). The concept of 'tubes of flow' implies that a tube—the region between any two neighbouring lines of flow—carries a uniform quantity of heat. The technique is to guess, and draw, the constant temperature surfaces, the equipotentials: then to draw the flow lines perpendicular to them; and finally, to measure the distance apart of the flow lines, and to calculate the varying resistance of a tube (as in electricity, α 1/cross-section) and finally to re-allocate the total temperature drop, according to the total resistance so far along the tube of flow. In principle, it would be possible to do this iteratively, re-drawing the equipotentials again and again; somehow we doubt if our grandfathers actually did this!

12.8   FINITE DIFFERENCE METHODS

Assuming you are an engineer, we should satisfy your hunger for numbers, very soon. In section 8.4, we saw the general idea of replacing $\nabla^2\theta$ by

$$\frac{\theta(x-h, y) + \theta(x+h, y) + \theta(x, y-h) + \theta(x, y+h) - 4\theta(x, y)}{h^2}$$

as in Fig. 12.4—the 'star' pattern. We shall now solve the problem $\nabla^2\theta = -2$, which would describe a plate sitting in the sun, with its boundaries cooled, as

$$q = \theta(x, y + h)$$

$$p = \theta(x - h, y) \qquad\qquad t = \theta(x, y) \qquad\qquad r = \theta(x + h, y)$$

$$s = \theta(x, y - h)$$

Fig. 12.4   The star pattern, representing the finite difference formula:

$$\nabla^2\theta = (p + q + r + s - 4t)/h^2$$

in Fig. 12.5. It is of technical importance in that the solution describes the torsion of a square rod. Using Fig. 12.4 with $p = q = r = s = 0$, Fig. 12.5(a) gives the single equation $-4a = h^2\nabla^2\theta = -2$ because $h = 1$; hence $a = \frac{1}{2}$.

In fact, we want to calculate the integral of θ, because this is given in textbooks. Consider

$$\int_{-1}^{1} \theta(x, y)\, dx.$$

At $y = \pm 1$ this is zero; at $y = 0$, it is 2/3, by Simpson's rule. Applying Simpson's rule in y, we have $\iint \theta\, dx\, dy = 2 \cdot \frac{2}{3} \cdot \frac{2}{3} = \frac{8}{9}$.

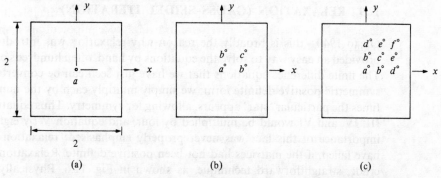

Fig. 12.5 θ is zero along the edges of this 2×2 square plate. Within the plate, $\nabla^2\theta = -2$. In (a) we have one variable, with $h = 1$. In (b), $h = \frac{1}{2}$, giving nine variables; by symmetry, this reduces to the three shown. In (c) we have $h = 1/3$, giving 16 variables reducing by symmetry to the six shown.

Moving on to Fig. 12.5(b), and applying Fig. 12.4, we have the equations

$$4b - 4a = -2 \times (\tfrac{1}{2})^2 = -\tfrac{1}{2} \text{ (using } a \text{ as centre of 'star', } h = \tfrac{1}{2}$$

$$2b - 4c = -\tfrac{1}{2} \text{ (using } c \text{ as centre of 'star' and remembering}$$
$$\text{boundary values are zero)}$$

$$2c + a - 4b = -\tfrac{1}{2} \text{ (using } b \text{ as centre of 'star')}$$

hence $a = \tfrac{9}{16}$, $b = \tfrac{7}{16}$, $c = \tfrac{11}{32}$. Using the $7 - 32 - 12 - 32 - 7$ rule in x for $y = 0$ gives $(32b + 12a + 32b)/45$, and for $y = \pm\tfrac{1}{2}$ gives $(32c + 12b + 32c)/45$. Applying the $7 - 32 - 12 - 32 - 7$ rule in y gives $\iint \theta \, dx \, dy = 1.067160$.

Moving on to Fig. 12.5(c) the equations are

$$4b - 4a = -2 \times (\tfrac{1}{3})^2 = -2/9 \dots \text{I}$$

$$a + 2c + d - 4b = -2/9 \dots \text{II}$$

$$2b + 2e - 4c = -2/9 \dots \text{III}$$

$$b + 2e - 4d = -2/9 \dots \text{IV}$$

$$d + c + f - 4e = -2/9 \dots \text{V}$$

$$2e - 4f = -2/9 \dots \text{VI}$$

giving $a = 15/26$, $b = 61/117$, $c = 17/36$, $d = 40/117$, $e = 73/234$, and $f = 11/52$. To find the integral, we apply the higher Newton–Cotes rule

$$\int_{-1}^{1} \phi(x) \, dx = \frac{2}{840} \{41\phi(-1) + 216(-\tfrac{2}{3}) + 27\phi(-\tfrac{1}{3}) + 272\phi(0)$$
$$+ 27\phi(\tfrac{1}{3}) + 216\phi(\tfrac{2}{3}) + 41\phi(1)\},$$

and it gives 1.0983. The accepted answer is 1.1246. Thus the errors are proportional to h^2. Therefore to get good answers we should have to solve a lot of equations!

12.9 RELAXATION (GAUSS–SEIDEL ITERATION)

On to 1940—this is broadly the reason why relaxation was introduced; it provided an easy way to solve the equations by hand, often hundreds of them. The finite difference equations that we have just seen can be converted into symmetric positive definite form; we simply multiply each by the number of times the particular 'star' appears, allowing for symmetry. Thus equations II, III, IV, and VI would be multiplied by four, and equation V by eight. The importance of this fact was never properly emphasised; relaxation would have failed, if the matrices had not been positive definite. Relaxation was a crude, straightforward technique, as shown in Fig. 12.6. Physically, there would be a large diagram on paper. We used to change the numbers on the nodes progressively, using an eraser, and iterate until they settled down.

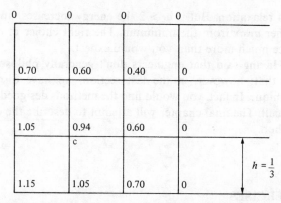

Fig. 12.6   With simple relaxation, we would correct the value $0.95 = c$ by writing the 'star' equation from Fig. 12.5, $0.60 + 0.60 + 1.05 + 1.05 - 4c = -4h^2$, giving $c = 0.94$. Then replacing 0.95 by 0.94, we would proceed to the next variable, and update it in the same way.

Even with a repertoire of tricks, like 'block relaxation'—altering many of the numbers simultaneously—a real example might take a week or more to solve. Nowadays, nobody does such calculations by hand, but one favourite trick 'over-relaxation', is still used on computers—not only in the context of relaxation. A single step is described in the caption to Fig. 12.6. The equation to be solved is

$$0.60 + 0.60 + 1.05 + 1.05 - 4c = -4h^2, \quad h = 1/3,$$

and the solution is

$$c = (4h^2 + 0.60 + 0.60 + 1.05 + 1.05)/4.$$

But with an 'over-relaxation factor' ω, we would replace 0.95 not by c but by $c^* = 0.95 + \omega(c - 0.95)$—i.e. we amplify the 'jump' by the factor ω. Typically $\omega = 1.75$ to 1.95 in practice, or even more. However, ω never exceeds 2, for the reason shown in Fig. 12.7. If $\omega < 2$, the energy decreases slightly with

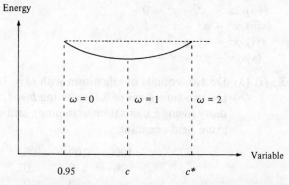

Fig. 12.7   In a positive definite problem, there is always an 'energy' which can be minimised, a quadratic function of the variables, in particular of the single variable currently being updated. Thus $\omega = 2$ restores the previous energy level, and $\omega > 2$ gives an increase in energy.

each relaxation. But if $\omega > 2$ the energy increases, and moves further and further *away* from the minimum. The right choice of ω accelerates convergence much more than you would expect.

Having said that engineers don't generally philosophise, we have done just that; however, in the last section, we have discussed an outmoded technique. In fact, you would find the methods designed for computers rather difficult. The final chapter will attempt to describe the essentials of one such method.

EXERCISES

12.1. Describe the following equations as hyperbolic, parabolic, or elliptic:

(a) $\dfrac{\partial^2 u}{\partial x^2} = \dfrac{\partial^2 u}{\partial t^2}$

(b) $\dfrac{\partial^2 u}{\partial x^2} = \dfrac{\partial u}{\partial t}$

(c) $\dfrac{\partial^2 u}{\partial x^2} + \dfrac{\partial^2 u}{\partial x \partial y} = 0$

(d) $\dfrac{\partial^2 u}{\partial x^2} + \dfrac{\partial^2 u}{\partial y^2} = 0.$

12.2. What curves, surfaces do the following represent?
(i) $xy = 1$
(ii) $(x + y)(x - y) = 1$
(iii) $x^2 + y^2 + \frac{1}{2}z^2 = 1$
(iv) $x^2 + y^2 = z$
(v) $2x^2 + 2xz + 2y^2 + 2yz + z^2 = 1$
(vi) $(x + y)^2 = 0$
(vii) $2x - x^2 - y^2 = 0$
(viii) $x^2 + xy - y^2 = 1$
(ix) $x^2 + xy + y^2 = 1$
(x) $(x - y)^2 = x + y$

12.3. (i) (a) Do two rounds of relaxation with $\omega = 1.5$.
(b) Do two time steps of 0.1 assuming $h = 1$, thermal properties of unity (using Euler's time-marching), and the boundary temperature held constant.

100	100	100	100
90	0	0	90
	80	0	80
		70	70

(ii) (a) Do two rounds of relaxation with $\omega = 4/3$.

(b) Do two time steps of 0.1 assuming $h = 1$, thermal properties of unity and boundary temperature held constant.

0	100	200	300
100	0	0	200
200	0	0	100
300	200	100	0

13

Splines and finite elements

13.1 WHITHER FINITE ELEMENTS?

For the moment, finite elements are widely used, mostly in stressing structures, but also to an increasing extent to predict the flow of oil underground,
etc. To some extent, it is the market supply directing the demand: the annual
usage and the expenditure on software runs to about a thousand million
dollars globally. Many sophisticated packages are readily available, and
strongly advertised.

Cubic splines are now a preferred way of interpolating. It is usual to
approach them algebraically, by matching the continuity requirements of the
polynomial responses in neighbouring sections. We choose instead an
approach that will serve as a good introduction to most of the basic
techniques of finite elements. It's a little harder. But we think you will
eventually be grateful to us.

We were disappointed in section 4.5 to see that Lagrange interpolation
gives silly answers near the end points. As promised, we now try a new
approach, and draw the curve using a springy beam, bent so that it passes
through the given points, as in Fig. 13.1. Many years ago, draughtsmen
actually used such beams in the office. Observe that beyond the ends, in
regions A*A and HH*, the curvature d^2w/dx^2 must be zero. Near the ends A
and H the curvature will tend to be small. This suggests that it behaves less
wildly—indeed near A where there should be curvature, the curve does not
provide it. It displays the opposite fault to Lagrange! In general, the piecewise
polynomial representations—as in finite elements—are preferred, because
they are less temperamental than a single overall polynomial.

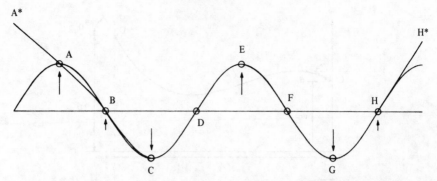

Fig. 13.1   This beam is forced to pass through points A–H by forces as shown. Therefore beyond these points, in segments AA* and HH*, the beam experiences no bending. Observe that the errors are bad in AB, but negligible in GH. Similar to our attempt in Fig. 4.4: interesting. (To get this picture, we interpolated cubic splines numerically onto a sine curve.)

In our beam theory, we must keep the slopes so small that the slope $dw/dx = \theta \cong \sin\theta$. Otherwise the theory would be much more difficult. Indeed, at this point we can say goodbye to structural theory specific to beams. Instead, we note that an unloaded beam remains straight; that the natural measure 'deformation', whether local or general, is the curvature. Furthermore, we assert that the deflection w is a cubic in x between any two successive points, and that

$$\frac{1}{2} \int_{A}^{Z} \left(\frac{d^2 w}{dx^2}\right)^2 dx = \phi = \text{minimum}.$$

i.e., regardless of sign, we try to reduce the curvature everywhere; the 'strain energy' in bending is a minimum. Our curve will then have the least possible overall 'wiggliness'.

13.2   SHAPE FUNCTIONS

The values $w_A \ldots w_H$ at the nodes A $\ldots$ H are given (Fig. 13.1). The values of the slopes $\theta_A \ldots \theta_H$ are chosen to minimise ϕ. Segments like AB, BC are called 'elements' for convenience. Segments A*A and HH* (Fig. 13.1) contribute nothing to ϕ. Element AB is called element I, BC is element II, CD is element III, etc., and each contributes something to ϕ. Thus it is natural to write ϕ separately for each segment like CD. We say

$$\phi = \phi^{\text{I}} + \phi^{\text{II}} + \cdots + \phi^{\text{VIII}}$$

Let's concentrate on element III = CD. In finite elements we always use local axes; here we measure x to the right from C. Fig. 13.2 introduces the idea of 'shape functions'. For example, the first shape function N_1 has unit value at C, but the slope at C and the value and slope at D are all zero. The shape functions are cubics. There are four variables, w_C, θ_C, w_D, and θ_D which

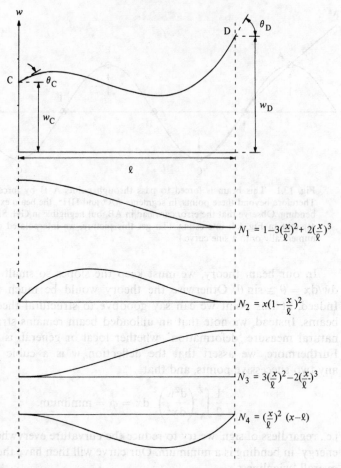

Fig. 13.2 The curve between C and D is a cubic in x, with four independent coefficients. The shape functions $N_1 \ldots N_4$ are also cubics. Therefore w may be written as $N_1 w_C + N_2 \theta_C + N_3 w_D + N_4 \theta_D$, because the shape functions have unit and zero values and slopes at C and D in the right combinations.

give us just enough information to define a cubic $p + qx + rx^2 + sx^3$ over the element; w_C times (function which gives unit w_C and zero in $\theta_C \ldots w_D$) plus θ_C times (function which gives unit θ_C and zero in the other three variables) plus ... plus ... must be a cubic with the correct values and slopes, w_C, θ_C, w_D, and θ_D. The precise way we handle this should remind you of Lagrange interpolation; indeed, that is why we called Lagrange's expressions 'shape functions'.

Future intentions. We called w_C, θ_C, w_D, and θ_D 'variables'. This was misleading; we are given w_C and w_D. The true variables are θ_C and θ_D, and θ_A and θ_B, too. We must write ϕ^I, $\phi^{II} \ldots \phi^{VIII}$ then differentiate with respect to $\theta_A \ldots \theta_H$ in turn, and solve the eight equations for $\theta_A \ldots \theta_H$. We need not differentiate with respect to $w_A \ldots w_H$ which we can't vary anyway.

13.3 ELEMENT STIFFNESS MATRICES

We now use the techniques of section 5.3:

$$\phi^{III} = \frac{1}{2} \int_0^l (w'')^2 \, dx = \frac{1}{2} \int_0^l (w_C N_1'' + \theta_C N_2'' + w_D N_3'' + \theta_D N_4'')^2 \, dx \quad (13.1)$$

Thus for example

$$\frac{\partial \phi^{III}}{\partial \theta_C} = \int_0^l N_2''(w_C N_1'' + \theta_C N_2'' + w_D N_3'' + \theta_D N_4'') \, dx.$$

Treating w_C, w_D, and θ_D similarly, and writing the four expressions in matrix form

$$\begin{bmatrix} \partial \phi^{III}/\partial w_C \\ \partial \phi^{III}/\partial \theta_C \\ \partial \phi^{III}/\partial w_D \\ \partial \phi^{III}/\partial \theta_D \end{bmatrix} = \int_0^l \begin{bmatrix} N_1'' \\ N_2'' \\ N_3'' \\ N_4'' \end{bmatrix} \begin{bmatrix} N_1'' & N_2'' & N_3'' & N_4'' \end{bmatrix} \begin{bmatrix} w_C \\ \theta_C \\ w_D \\ \theta_D \end{bmatrix} dx$$

$$= \begin{bmatrix} \int N_1''^2 \, dx & \int N_1'' N_2'' \, dx & \int N_1'' N_3'' \, dx & \int N_1'' N_4'' \, dx \\ & \int N_2''^2 \, dx & \int N_2'' N_3'' \, dx & \int N_2'' N_4'' \, dx \\ & & \int N_3''^2 \, dx & \int N_3'' N_4'' \, dx \\ \text{Symm} & & & \int N_4''^2 \, dx \end{bmatrix} \begin{bmatrix} w_C \\ \theta_C \\ w_D \\ \theta_D \end{bmatrix} \quad (13.2)$$

because $w_C \ldots \theta_D$, although they are variables, do not vary during the integration.

The 4×4 matrix of coefficients in (13.2) is called the 'stiffness matrix', K, because traditionally almost the only revenue-earning tasks of finite elements have been in structures, where forces cause deflections, and (stiffness) = (force)/(deflection). It is equally valid, however, to regard $(K)^{III}$ as simply the 'Hessian' matrix of second derivatives of ϕ^{III} with respect to $w_C \ldots \theta_D$

$$K^{III} = \begin{bmatrix} \dfrac{\partial^2 \phi^{III}}{\partial w_C^2} & \dfrac{\partial^2 \phi^{III}}{\partial w_C \, \partial \theta_C} & \dfrac{\partial^2 \phi^{III}}{\partial w_C \, \partial w_D} & \dfrac{\partial^2 \phi^{III}}{\partial w_C \, \partial \theta_D} \\ \dfrac{\partial^2 \phi^{III}}{\partial \theta_C \, \partial w_C} & & \text{etc. (symm. matrix)} \\ \vdots & & \end{bmatrix}$$

For you can easily check that the second derivatives of (13.1) are the integrals comprising K in (13.2). Thus, for example,

$$K_{34} = \frac{\partial^2 \phi}{\partial w_D \, \partial \theta_D} = \int_0^l N_3'' N_4'' \, dx$$

$$= \int_0^l \left(\frac{6}{l^2} - \frac{12x}{l^3}\right)\left(\frac{6x}{l^2} - \frac{2}{l}\right) dx = -\frac{6}{l^2}$$

and the entire matrix is

$$K^{III} = \frac{1}{l^3} \begin{bmatrix} 12 & 6l & -12 & 6l \\ 6l & 4l^2 & -6l & 2l^2 \\ -12 & -6l & 12 & -6l \\ 6l & 2l^2 & -6l & 4l^2 \end{bmatrix}.$$

In generating the much larger **K** matrices to be found in commercial finite element packages, typically 12×12 up to 60×60, this step is often done using a Gauss rule. However, this 'beam' stiffness is universally available.

13.4   FINITE ELEMENTS: ASSEMBLY OF EQUATIONS

There are two complications. The first is that so far we have been looking at the contributions of ϕ^{III} alone. We have to get used to the idea of writing the contributions to the equations, the derivatives of ϕ, also from ϕ^{III} alone. This is pointless in itself; it is only when we bring the component elements together that the exercise is of any practical use. The second complication is that w_C and w_D are given, so that we must ignore $\partial \phi / \partial w_C$ and $\partial \phi / \partial w_D$, equations 1 and 3 of (13.2), i.e. rows 1 and 3 of **K**; substituting from **K**, rows 2 and 4 give $\partial \phi / \partial \theta_C = \partial \phi / \partial \theta_D = 0$:

$$\frac{1}{l^3} (6lw_C + 4l^2\theta_C - 6lw_D + 2l^2\theta_D) = 0$$

$$\frac{1}{l^3} (6lw_C + 2l^2\theta_C - 6lw_D + 4l^2\theta_D) = 0,$$

or, taking the constant w terms to the right-hand side,

$$\frac{1}{l}\begin{bmatrix} 4 & 2 \\ 2 & 4 \end{bmatrix}\begin{bmatrix} \theta_C \\ \theta_D \end{bmatrix} = \frac{6}{l^2}\begin{bmatrix} w_D - w_C \\ w_D - w_C \end{bmatrix}.$$

Solving these equations would give the response of element III alone. So it is reassuring to find that if we have an isolated element, then $\theta_C = \theta_D = (w_D - w_C)/l$, i.e. the beam CD remains straight.

This is of no use to anybody, and we must combine the elements, i.e. consider $\phi = \phi_I + \phi_{II} + \phi_{III}$. We must also write the derivatives with respect to $\theta_A \ldots \theta_D$. Element I, for example, extends from A to B, so that A replaces C, B replaces D. Thus combining the coefficients and the right-hand sides from $\phi^I + \phi^{II} + \phi^{III} = \phi$, just three elements.

$$
\begin{bmatrix}
\boxed{\begin{matrix} 4/l_\mathrm{I} & 2/l_\mathrm{I} \\ 2/l_\mathrm{I} & 4/l_\mathrm{I} \end{matrix}} + \begin{matrix} \text{element II} \\ \boxed{\begin{matrix} 4/l_\mathrm{II} & 2/l_\mathrm{II} \\ 2/l_\mathrm{II} & 4/l_\mathrm{II} \end{matrix}} \end{matrix} + \begin{matrix} \text{element III} \\ \boxed{\begin{matrix} 4/l_\mathrm{III} & 2/l_\mathrm{III} \\ 2/l_\mathrm{III} & 4/l_\mathrm{III} \end{matrix}} \end{matrix}
\end{bmatrix}
\begin{bmatrix} \theta_A \\ \theta_B \\ \theta_C \\ \theta_D \end{bmatrix}
$$

$$
= \begin{bmatrix}
6(w_B - w_A)/l_\mathrm{I}^2 \\
6(w_B - w_A)/l_\mathrm{I}^2 + 6(w_C - w_B)/l_\mathrm{II}^2 \\
6(w_C - w_B)/l_\mathrm{II}^2 + 6(w_D - w_C)/l_\mathrm{III}^2 \\
6(w_D - w_C)/l_\mathrm{III}^2
\end{bmatrix}
$$

Of course, this blending of the three sets of equations is very simple here, not at all typical of real finite elements. When we combine the contributions from many triangles and quadrilaterals in 2D, or bricks in 3D, it is seldom possible to 'frame' the individual contributions from the element **K** matrices, as we have done above using dotted lines. In the case of Fig. 13.1, we have seven elements (of equal length) and eight slopes, giving the eight equations:

$$
\begin{bmatrix}
4 & 2 & & & & & & \\
2 & 8 & 2 & & & & & \\
& 2 & 8 & 2 & \text{zero} & & & \\
& & 2 & 8 & 2 & & & \\
& & & 2 & 8 & 2 & & \\
\text{zero} & & & & 2 & 8 & 2 & \\
& & & & & 2 & 8 & 2 \\
& & & & & & 2 & 4
\end{bmatrix}
\begin{bmatrix} \theta_1 \\ \theta_2 \\ \theta_3 \\ \theta_4 \\ \theta_5 \\ \theta_6 \\ \theta_7 \\ \theta_8 \end{bmatrix}
=
\begin{bmatrix} -6 \\ -12 \\ 0 \\ 12 \\ 0 \\ -12 \\ 0 \\ 6 \end{bmatrix}.
$$

To draw the interpolated curve in Fig. 13.1 we take the θ values from these equations, and the given values w, and multiply into the shape functions, which are functions of x within each element. This is easy to program, see SPLINE.

13.5 EXPERIENCE, AND QUINTIC SPLINES

You should now use the program experimentally, to learn what happens in practice. (Note: we have defined the function as $f(z) = \cos(z)$, but you should change it, just to see how splines work with different functions.) This is an unfortunate phase in the course. We tell you—and it is true—that cubic splines are very widely used. We tell you, that the reason for using cubic splines rather than Lagrange interpolation is that crazy things are less likely to happen near the ends. Then you do an example, and the results are something like Fig. 13.1, very bad near the end that requires curvature. Yes, we are probably right if you try to extrapolate, if you compare the curves beyond the endpoint.

The fact is, probably, that your Lagrange is very much better than the spline in the middle of the range, and near the ends is not much worse. But we are not comparing like with like. If we have six points, Lagrange gives a quintic; and the end elements in the spline really only have three degrees of freedom, since the end slope takes its 'relaxed' value. In fact, with many points the cubic spline is much cheaper in arithmetic than Lagrange interpolation, although the theory of cubic splines is more difficult.

We need not limit ourselves to cubic splines: there are, for example, quintic splines, with w given at each node, and w' and w'' as variables. This formulation minimises

$$\phi = \frac{1}{2} \int_A^Z (w''')^2 \, dx.$$

If you write a program for quintic splines, and play games with it, you may discover two interesting things about it:

(i)   If you have only two or three points, i.e. one or two elements, the computer finds that it cannot solve the equations. There is a zero pivot; the assembly matrix is singular.

(ii) You expect the curvature to be continuous between elements, because at the node between them w'' is a variable. But in fact there is continuity of the fourth derivatives.

When quintic splines work well, we would challenge anybody to guess, by looking at the resulting curve, where the given points are; the curve is so incredibly smooth. The same thing happened in the cubic spline, but less spectacularly; anybody who remembers a little about beams and about statics, will recall that the bending moments are continuous from each segment to the next: i.e. w'' is continuous, whereas the nodal variables only guarantee that the slopes w' are continuous. With quintic splines, derivatives up to the fourth are matched between elements. In mathematical parlance, cubic splines give $C^{(2)}$ continuity; quintic splines give $C^{(4)}$ continuity.

There is a more subtle reason for the remarkably good performance of quintic splines. Experience with finite elements shows that if the formulation is in some way close to a case that gives unsolvable equations, then the performance seems to benefit, usually. With quintic splines, you need at least three elements to avoid a singular matrix.

Why then aren't quintic splines fashionable? Our guess is that, without finite elements, quintic splines would be too complicated. People are unaware of how good they are. Interactions between disciplines can be very helpful!

Anybody intrigued by these comments should read our *Finite Element Primer*. Splines are an excellent introduction, because they have taught you that it is better to break your problem into many regions, with a different but simple function in each. It may be cheaper, but not necessarily. It tends to be more reliable. This is why people like to use finite elements, even for problems which are not 'elliptic' in the strict sense.

EXERCISES

13.1. Using cubic splines, determine from the data

(i) $f(2.5)$ for:

x	1	2	3	4	5
$f(x)$	5.0	7.0	5.0	4.0	6.0

(Note the x values are evenly spaced.)

(ii) $f(15)$ for:

x	2	5	9	12	17
$f(x)$	2.0	4.0	3.0	3.0	8.0

(Note the x values are *not* evenly spaced.)

Check your answers: (sketch?, Lagrange?).

13.2. Determine the values of A, B, C, and D so that $f(x)$ is a cubic spline and that $\int_0^2 (f''(x))^2 \, dx$ is a minimum, where

$$f(x) = \begin{cases} 6 + 3x - 8x^3 & 0 \leq x \leq 1 \\ A + B(x-1) + C(x-1)^2 + D(x-1)^3 & 1 \leq x \leq 2 \end{cases}.$$

Leonora

Confrontations III v 1–16

1. Verily I say unto thee, ye compute for insight, thy numbers are but chaff.
2. There is naught sacred concerning numbers that issue from a computer.
3. Garbage in garbage out, thine algorithms do but confuse all the people.
4. Thou doest wrong to burn thy garbage. Desist ye, pollute not the air.
5. The virtue of numbers surpasseth not the thought that lieth behind them.
6. Treat ye all numbers with dire suspicion and keep ye from unclean thoughts.
7. For the wages of iniquity are to repeat thy numberings many times over.
8. Beware of lies, damned lies, and statistics, trust ye not in vanities unchecked.
9. The fowl of the air do drop forth droppings; likewise are the fruits of the computer.
10. Be not ashamed to confess thine abhorrent faults, for the computer knows no shame.
11. Thy computer thinketh not, but obeyeth instructions from whence they come.
12. Programs cometh not from Above but are conceived in manifest sin.
13. Yea, even an event of small probability happeneth on occasion, by merest chance.
14. Bear ye not witness against thy brother's statistics lest he pierce thine own.
15. Compute not the statistics of they neighbour's wife, there are statistics enough.
16. The beasts of the field are wise for they compute not, neither do they blaspheme.

Appendix

The numerical methods programs in this appendix were originally written by Bruce Irons. They have been updated to their current form by Jan Irons. The graphics routines are entirely attributable to her.

Here are the programs that we refer to in the main text:

- Program PERSPE calculates points for plotting in perspective projection.

- Program LAGRAN programs a Lagrange interpolation.

- Program LEASQ finds the least squares fitting with a polynomial through random points.

- Program CURFIT finds the least squares fitting of a given function by sampling the function at its Gauss points.

  (LEASQ and CURFIT) both refer to the subroutines GAUSS and VECTOR. These subroutines follow the two programs.)

- Program THERM calculates the heat transient of a metal bar initially at temperature zero.

- Program HEUN integrates by Heun's method.

- Program RUNGE integrates by Runge's method.

- Program POSIN inverts a positive definite symetric matrix by part inversion.

- Program EIG calculates the eigenvalues of a matrix.

- Program RELAX introduces Gauss-Seidel relaxation methods.

- Program SPLINE calculates a spline interpolation.

You will notice that there are characters '%' and '!' which appear in column 73 of the files. DO NOT copy these characters! They signal respectively:
%: the program line is relevant to graphics output only.
!: the program line is relevant to text output only.

These programs are accompanied by a library of graphics routines.

The first file contains routines that are independent of the plotting device used. i.e. they are portable from one system to another. These subroutines are:

- Subroutine AXIS draws the axes for a new plot and calculates the scale.

- Subroutine SCALER called by AXIS calculates the scale.

- Subroutine TRACE draws a line using the scale calculated in AXIS.

- Subroutine CROSS draws a cross.

The second file contains primitive graphic routines that must be changed for each device used. The example file is written in CALCOMP commands. However we have successfully run these programs using other graphics software. These subroutines are:

- Subroutine GRAFON turns on the graphics device and initializes variables.

- Subroutine GRAFOF turns off the graphics device.

- Subroutine NEWGRF readies the graphics device for a new plot.

- Subroutine DRAW moves the pen (up or down, absolute of relative).

- Subroutine RITER writes a string.

```
C*** Program to calculate points for plotting in perspective projection.
C*** The points are read from a user file. They must be entered one per
C*** line.
C
      PROGRAM PERSPE
C
C     POINTS are the global object coordinates (vector Q in notes).
C     EYEVEC gives the global eye coordinates (vector E in notes).
C     VECSTA are the object coordinates when the origin is shifted to the
C     eye position.
C     XYZ are the unit viewing vectors.
C     XZPLOT will contain the converted object coordinates.
C     Note that we predifine the unit file numbers IIFILE and IOFILE to
C     assure portability.
C
      DIMENSION POINTS(3,100), XXZPLOT(2,100), EYEVEC(3), VERT(3), VECSTA(3,100),
     .XYZ(3,3)
      CHARACTER*72 FILNAM
      CHARACTER TITLE*66
      DATA VERT /1.0E-12, 0., 1./
      DATA IIFILE /5/
      DATA IOFILE /6/
C
C*** The file OUTPUT.DAT is created for program output.
C
      OPEN(IOFILE, FILE='OUTPUT.DAT', STATUS='NEW', ERR=1,
     .ACCESS='SEQUENTIAL')
      GOTO2
1     OPEN(IOFILE, FILE='OUTPUT.DAT', STATUS='OLD',
     .ACCESS='SEQUENTIAL')
2     CONTINUE
C
C*** Collect all the geometric data required.
C
```

```fortran
      WRITE (*,*) 'Name of file of points ? (Type ''Q'' to quit) '
      READ (*,'(A)') FILNAM
      IF (FILNAM.EQ.'Q ') THEN
.
      CLOSE (IOFILE, STATUS='KEEP')
      STOP
      ENDIF
      OPEN(IIFILE, FILE=FILNAM, STATUS='OLD', ACCESS='SEQUENTIAL',
     & ERR=5)
      GOTO 6
5     WRITE (*,*) 'Error opening file ', FILNAM
      CLOSE (IOFILE, STATUS='KEEP')
      STOP
6     CONTINUE
      DO 10 NP = 1,101
      IF (NP.GT.100) THEN
         WRITE (*,*) 'Only first 100 points can be processed.'
      ELSE
         READ (IIFILE, *, END=20) (POINTS(I,NP), I = 1,3)
      ENDIF
10    CONTINUE
20    NPOINT = NP - 1
      IF (NPOINT.LE.1) THEN
         WRITE (*,*) 'At least two points are needed.'
         CLOSE (IOFILE, STATUS='KEEP')
         STOP
      ENDIF
C
      CALL GRAFON
C
30    WRITE (*,*) 'What is the viewing distance ? (Enter 0. to quit)'
      READ (*,*) DIST
      IF (ABS(DIST).LT.1.E-12) GOTO 1000
      IF (DIST.LE.0.) THEN
         WRITE (*,*) 'Viewing distance must be positive. Try again. '
         GOTO 30
```

```
        ENDIF
        WRITE (*,*) 'Now give the x,y,z for the eye coordinates ?'
        READ (*,*) (EYEVEC(I),I=1,3)
        WRITE (IOFILE,'(A,F8.4)') ' Viewing distance = ',DIST
        WRITE (IOFILE,'(A,3(F8.4,A))') 'Eye position = (',
       . EYEVEC(1),',',EYEVEC(2),',',EYEVEC(3),')'
C
C***    Calculate the coordinates relative to given eye-position.
C***    Also the centroid relative to these points for initial viewing
C***    position.
C
        DO 80 NP = 1,NPOINT
        DO 80 I = 1,3
        VECSTA(I,NP) = POINTS(I,NP) - EYEVEC(I)
        XYZ(I,2) = XYZ(I,2) + VECSTA(I,NP)
 80     CONTINUE
C
C***    Calculate the unit viewing vectors.
C
        DO 160 ITER = 1,3
        CALL VECTOR(XYZ(1,2), VERT(1,1))
        CALL VECTOR(XYZ(1,1), XYZ(1,3))
        DO 100 J = 1,3
        SCAL = SQRT(SCALAR(XYZ(1,J),XYZ(1,J)))
        IF (SCAL.EQ.0.) THEN
        WRITE (*,*) 'Eye is at centroid of object.'
        GOTO 1000
        ENDIF
        DO 100 I = 1,3
        XYZ(I,J) = XYZ(I,J)/SCAL
 100    CONTINUE
        XMIN = 1.0E12
        XMAX = -XMIN
        ZMIN = XMIN
        ZMAX = XMAX
C
C**     Do the perspective transformation.
C
```

```fortran
      DO 120 NP = 1,NPOINT
      DEN = SCALAR(VECSTA(1,NP), XYZ(1,2))/DIST
      IF (ABS(DEN).LT.1.0E-12) DEN = SIGN(1.0E-12,DEN)
      X = SCALAR(VECSTA(1,NP), XYZ(1,1))/DEN
      Z = SCALAR(VECSTA(1,NP), XYZ(1,3))/DEN
      IF(X.LT.XMIN) XMIN = X
      IF(X.GT.XMAX) XMAX = X
      IF(Z.LT.ZMIN) ZMIN = Z
      IF(Z.GT.ZMAX) ZMAX = Z
      XZPLOT(1,NP) = X
      XZPLOT(2,NP) = Z
  120 CONTINUE
C*** Make plot of points.
C
      IF (ITER.EQ.3) THEN
      WRITE (TITLE,'(A,3(F7.3,A))')
     . 'Perspective projection from eye position (', EYEVEC(1),
     . ',', EYEVEC(2), ',', EYEVEC(3), ')'
      CALL AXIS(XMIN, XMAX, ZMIN, ZMAX, 'x-axis', 'z-axis', TITLE,
     . 6, 66)
      ILINE = 0
      DO 125 NP = 1,NPOINT
      CALL TRACE (XZPLOT(1,NP), XZPLOT(2,NP), ILINE)
      CALL CROSS
      ILINE = 1
  125 CONTINUE
C*** Write data to file
C
      WRITE (6, '(//5X,A/)')
     . 'PERSPECTIVE CONVERSION OF OBJECT COORDINATES'
      DO 130 I = 1,NPOINT
      WRITE (6, '(A,6(F8.3,A))')
     . '(', POINTS(1,I), ',', POINTS(2,I), ',', POINTS(3,I),
     . ') ===> (', XZPLOT(1,I), ',', XZPLOT(2,I), ')'
```

```fortran
130   CONTINUE
      ELSE
C*** Now get improved approximations for the optimum viewing direction.
C
      DO 140 I = 1,3
        XYZ(I,2) = XYZ(I,2)*DIST + XYZ(I,1)*(XMIN+XMAX)/2
     +  + XYZ(I,3)*(ZMIN+ZMAX)/2
140   CONTINUE
      ENDIF
160   CONTINUE
C*** Do diagnostics for unsuitable eye positions.
C
      IF(XMAX/DIST.GT.0.4 .OR. ZMAX/DIST.GT.0.4) THEN
        WRITE(*,*)
     +  'You are too close to the viewing object. Try again (Y/N) ?'
      ELSE IF(XMAX/DIST.GT.0.1 .OR. ZMAX/DIST.GT.0.1) GOTO 30
        WRITE(*,*)
     +  'You are too far from the viewing object. Try again (Y/N) ?'
      ENDIF
      READ (*,'(A)') FILNAM
      IF(FILNAM.EQ.'Y') GOTO 30
C
1000  CALL GRAFOF
      CLOSE(IOFILE, STATUS='KEEP')
      STOP
      END
```

```
C
C***  To calculate the vector cross product:  W = U x V.
C
      SUBROUTINE VECTOR(U, V, W)
      DIMENSION U(3), V(3), W(3)
      K = 3
      DO 2 I = 1,3
      W(6-I-K) = U(K)*V(I) - V(K)*U(I)
      K = I
    2 CONTINUE
      RETURN
      END
C
C***  To calculate the scalar product of two vectors: U . W.
C
      FUNCTION SCALAR(U,V)
      DIMENSION U(3), V(3)
      SCAL = 0.0
      DO 2 I = 1,3
      SCAL = SCAL + U(I)*V(I)
    2 CONTINUE
      SCALAR = SCAL
      RETURN
      END
```

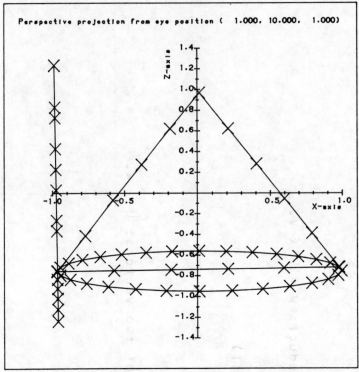

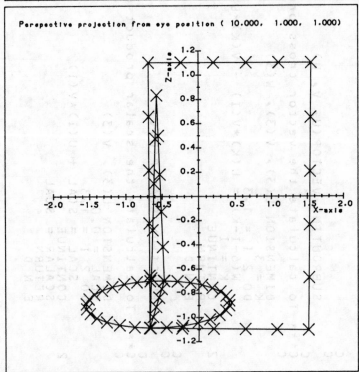

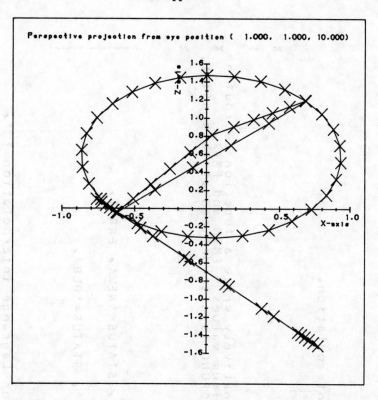

Perspective projection from eye position (1.000,   1.000, 10.000)

Different perspective views of a rectangle, a triangle and a circle in three different planes.

```fortran
C*** Teaching program for Lagrange interpolation.
C
      PROGRAM LAGRAN
      DIMENSION XGIV(20), FGIV(20)
      LOGICAL LFIRST
      CHARACTER POINT*7
      DATA IOFILE /6/
C
C*** Define the function FUNC. Alternatively write a function subroutine %
C*** Also define its minimum and maximum values - FMIN and FMAX.         %
C*** The derivative of the function DFUNC is also defined at this point. %
C
      DATA FMIN, FMAX /-1., 1./
      FUNC(Z) = COS(Z)
      DFUNC(Z) = - SIN(Z)
C
C*** Open data output file.
C
      OPEN(IOFILE, FILE='OUTPUT.DAT', STATUS='NEW', ERR=1,
     .     ACCESS='SEQUENTIAL')
      GOTO 2
    1 OPEN(IOFILE, FILE='OUTPUT.DAT', STATUS='OLD',
     .     ACCESS='SEQUENTIAL')
    2 CONTINUE
C
C*** Read the points.
C
   20 WRITE (*,*) 'How many points for Lagrange interpolation ? '
     .           '(between 3 and 20)'
      READ (*,*) N
      IF(N.GT.20 .OR. N.LT.3) GOTO 20
      BIG = 0.
      WRITE (*,*) 'Enter X-values in increasing order.'
      DO 80 I = 1,N
```

```
60    WRITE(*,'(A,I2,A)') ' X Sample point ',I,' ?'
      READ(*,*) XGIV(I)
      IF(I.NE.1) THEN
        DEL = XGIV(I) - XGIV(I-1)
        IF(DEL.GT.BIG) BIG = DEL
        IF(DEL.LE.0.) THEN
          WRITE(*,*) ' X-values must be increasing order. Try again.'
          GOTO 60
        ENDIF
      ENDIF
      FGIV(I) = FUNC(XGIV(I))
80    CONTINUE
      START = XGIV(1) - 2.*BIG
      FIN = XGIV(N) + 2.*BIG
C
C***  Decide on a steplength for the table.
C
100   WIDTH = FIN-START
      WRITE(*,'(A/A/2(F8.5,A))')
     . ' What is the steplength in X ?',
     . ' (Value must be between ',WIDTH/300,' and ',WIDTH/30,')'
      READ(*,*) STEP
      NSTEP = IFIX(WIDTH/STEP + 1.1)
      IF(NSTEP.GT.301 .OR. NSTEP.LE.30) GOTO 100
      INT = 0
C
      LFIRST = .TRUE.
      CALL GRAFON
      FMAX = FMAX - FMIN
      FMAX = FMAX + 0.3*F
      FMIN = FMIN - 0.3*F
      CALL AXIS(START,FIN,FMAX,FMIN,'X-axis','Y-axis',
     . 'Lagrange interpolation of Points',6,6,32)
C
      FMAX = FMAX + 0.6*F
      FMIN = FMIN - 0.6*F
```

```
      WRITE(IOFILE,*)
     .'EXPERIMENTS WITH POLYNOMIAL LAGRANGE INTERPOLATION'
      WRITE(IOFILE,'(//A,(7F10.6/17X))')
     .'Given x-values:',(XGIV(I),I=1,N)
      WRITE(IOFILE,'(//14X,A,2X,A,4X,A,3X,A,1X,A,3X,A,2X,A)')
     .'X','Y approx','Y real','Y Error','Y approx','Y real',
     .'Y-Error'
C
C*** Do the Lagrange interpolation.
C
      DO 200 NST = 1,NSTEP
      X = START + STEP*FLOAT(NST-1)
      VALUE = 0.0
      DVALUE = 0.0
      DO 160 I = 1,N
      TERM = FGIV(I)
      XI = XGIV(I)
      FACT = 1.
      DO 140 J = 1,N
      IF(J.EQ.I) GO TO 140
      TOP = X - XGIV(J)
      IF(ABS(TOP).LE.1.E-12) TOP = SIGN(1.E-12,TOP)
      TERM = TERM*TOP/(XI-XGIV(J))
      FACT = FACT + 1./TOP
140   CONTINUE
      VALUE = VALUE + TERM
      DVALUE = DVALUE + TERM*FACT
160   CONTINUE
C
C*** ...and print the answers for comparison.
C
      FUN = FUNC(X)
      DFUN = DFUNC(X)
      ERROR = VALUE-FUN
      DERROR = DVALUE-DFUN
```

```fortran
C***
C*** Signal if X is near a sample point.
C
180   POINT = .
      IF(INT.NE.N) THEN
        IF(ABS(X-XGIV(INT+1)).LT. 0.50001*STEP) THEN
          INT = INT + 1
          POINT = .*POINT*.
          GOTO 180
        ENDIF
      ENDIF
      WRITE(IOFILE,'(1X,A,F7.3,2(2(1X,F9.4),1X,E9.3))')
     .  POINT,X,VALUE,FUN,ERROR,DVALUE,DFUN,DERROR
      IF(.NOT.FIRST) THEN
        CALL TRACE(X, FUN, SFUN, 0)
      ENDIF
      IF(POINT.EQ.'*POINT*') CALL CROSS
      IF(SVALUE.GT.FMIN .AND. SVALUE.LT.FMAX) THEN
        CALL TRACE(SX, SVALUE, 0)
        CALL TRACE(X, VALUE, 1)
      ENDIF
      ENDIF
      SFUN = FUN
      SVALUE = VALUE
      FIRST = .FALSE.
      SX = X
200   CONTINUE
      CLOSE(IOFILE, STATUS='KEEP')
      CALL GRAFOF
      STOP
      END
```

```fortran
C***   Teaching program for least squares fitting of a polynomial through
C***   random points.
C
       PROGRAM LEASQ
C
       DOUBLEPRECISION EQ(5,5), R(5), XGIV(20), YGIV(20)
       DIMENSION VEC(5)
       LOGICAL LCRASH
       CHARACTER POINT*7
C
C***   Polynomial fitting will be of degree NTERM.
C***   (NTERM less than or equal to 20))
C
       DATA NTERM /3/,
      .     IOFILE /6/
C
C***   Initialize variables.
C
       DO 11 N = 1,NTERM
          R(N) = 0.
          DO 11 I = 1,NTERM
             EQ(N,I) = 0.
   11  CONTINUE
C
C***   Read the number of points.
C
       OPEN (IOFILE,FILE='OUTPUT.DAT', STATUS='NEW', ERR=1,
      .     ACCESS='SEQUENTIAL')
       GOTO 2
    1  OPEN (IOFILE, FILE='OUTPUT.DAT', STATUS='OLD',
      .     ACCESS='SEQUENTIAL')
    2  CONTINUE
       WRITE (IOFILE,'(A//A/)')
      .' LEAST SQUARES FITTING OF A POLYNOMIAL THROUGH RANDOM POINTS.',
      .' Given points:'
```

```fortran
20    WRITE(*,'(A,I2,A)')
     .'How many points (X,Y) ? (Between ',NTERM,' and 20)'
      READ(*,*)NPOINT
      IF(NPOINT.LT.NTERM .OR. NPOINT.GT.20) GOTO 20
C
C***  Read the points.
C
      YMAX = -.1E-5
      YMIN = -YMAX
      WRITE (*,*) 'X-values must be given in increasing order.'
      DO 60 N = 1,NPOINT
      WRITE (*,'(A,I2,A)') 'Give X and Y coordinates for point ',N
25    READ (*,*) XGIV(N), YGIV(N)
      WRITE(IOFILE,'(2(A,F9.4,A)') ' ', XGIV(N), ' ',
      IF(N.GT.1) THEN
       IF(XGIV(N).LE.XGIV(N-1)) THEN
        WRITE(*,*)
      .'Points must have increasing X-values. Try again.'
        GOTO 25
       ENDIF
      ENDIF
C
      IF (YMAX.LT.YGIV(N)) YMAX = YGIV(N)
      IF (YMIN.GT.YGIV(N)) YMIN = YGIV(N)
C
C***  And create equations (upper triangle only) while points are being read.
C
      X = XGIV(N)
      Y = YGIV(N)
      CALL VECTOR (NTERM, VEC, X)
      DO 60 I = 1,NTERM
      V = VEC(I)
      R(I) = R(I) + Y*V
      DO 60 J = I,NTERM
      EQ(I,J) = EQ(I,J) + V*VEC(J)
```

```
60    CONTINUE
C***  Gaussian reduction.
C
      CALL GAUSS (EQ, R, NTERM, LCRASH)
      IF (LCRASH) THEN
        CLOSE (IOFILE, STATUS='KEEP')
        STOP
      ENDIF
C
C***  Prepare for final table.
C
      BIG = XGIV(NPOINT)
      SMALL = XGIV(1)
      CALL PLAY (SMALL, BIG, 0.1)
180   WIDTH = (A/A/2(F8.4,A)))
      WRITE(*,*)'What steplength would you like for the table ?'
      . '(Between ',WIDTH/300.,' and ',WIDTH/30.,')'
      READ(*,*) STEP
      NSTEP = IFIX(WIDTH / STEP + 1.1)
      IF (NSTEP.GT.301 .OR. NSTEP.LE.30.) GOTO 180
C
C***  Print the answers for comparison.
C***  Signal if x is near a sample point.
C
      WRITE (IOFILE,'(//14X,A,2X,A,3X,A,5X,A)')
      . 'X', 'Y-approx.', 'Y-given', 'Y-error'
      CALL PLAYFON(YMIN, YMAX, 0.3)
      CALL GRAFON (SMALL, BIG, YMIN, YMAX, 'X-axis', 'Y-axis'
      CALL AXIS (SMALL, BIG, YMIN, YMAX, 'X-axis', 'Y-axis'
      . 'Least squares fitting of a polynomial through random points'
      CALL PLAY (YMIN, YMAX, 0.5)
      ILINE = 0
      X = SMALL
      INT = 0
```

```fortran
C
      DO 320 NS = 1,NSTEP
      CALL VECTOR(NTERM, VEC, X)
      VALUE = 0.0
      DO 240 N = 1,NTERM
      VALUE = VALUE + R(N)*VEC(N)
240   CONTINUE
C
      POINT = ' '
      IF(INT.NE.NPOINT) THEN
      IF(ABS(X-XGIV(INT+1)).LT.0.50001*STEP)
     .        THEN POINT='*POINT*'
      POINT = INT + 1
      GOTO 250
      ENDIF
      ENDIF
250   IF(VALUE.LT.YMAX .AND. VALUE.GT.YMIN)
      IF CALL TRACE (X, VALUE, ILINE)
      ILINE=
      IF(POINT.EQ.'*POINT*') THEN
      WRITE(IOFILE,'(1X,A,F7.3,2(1X,F9.4),1X,E9.3)')
     .        POINT, X, VALUE, YGIV(INT), YGIV(INT)-VALUE
      ELSE
      WRITE (IOFILE,'(8X,F7.3,1X,F9.4)') X, VALUE
      ENDIF
      X = X + STEP
320   CONTINUE
C*** Write the crosses on the graph.
C
      DO 330 I=1,NPOINT
      CALL TRACE (XGIV(I), YGIV(I), 0)
      CALL CROSS
```

```
330   CONTINUE

      CLOSE (IOFILE, STATUS='KEEP')
      CALL GRAFOF
      STOP
      END

C
C     SUBROUTINE PLAY (AMIN, AMAX, APLAY)
C
C*** Subroutine to recalculate AMIN and AMAX with a margin of security APLAY.
C
      RANGE = APLAY * (AMAX - AMIN)
      AMIN = AMIN - RANGE
      AMAX = AMAX + RANGE
      RETURN
      END
```

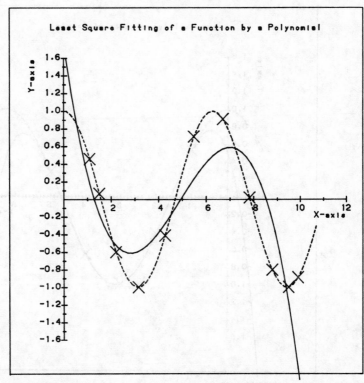

Least Square Fitting of a Function by a Polynomial

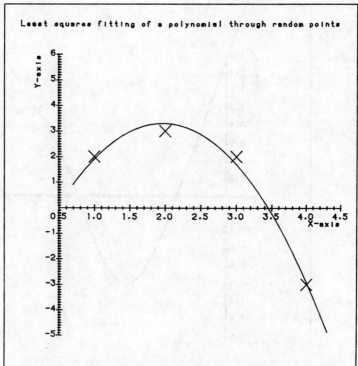

Least squares fitting of a polynomial through random points

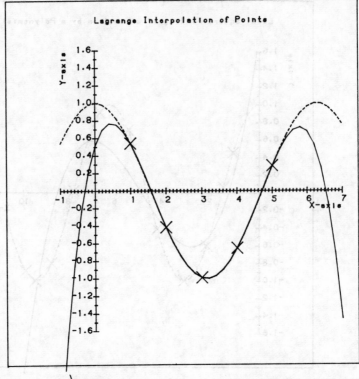

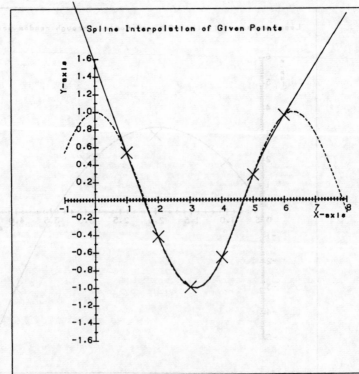

```
C*** Teaching program for curve-fitting a given function with a
C*** polynomial.
C
C*** NPOINT is the number of Gauss sample points. Note that XGAUSS and
C*** CGAUSS must be redefined for a change in NPOINT.
C*** XGAUSS contains the 11 Gauss sample points for an integral in
C*** C[-1,1].
C*** CGAUSS contains the weights of these sample values.
C*** NTERM-1 = the degree of the polynomial.
C*** (NTERM must be less than 6)
C
C       PROGRAM CURFIT
C
        LOGICAL LFIRST
        LOGICAL LCRASH
        DOUBLEPRECISION EQ(5,5), R(5)
        DIMENSION VEC(5), XGAUS(11), CGAUS(11)
        DIMENSION DR(5)
        DATA XGAUS/-.97822866, -.88706260,
       .           -.73015201, -.51909613,
       .           -.26958657, -.0125580 37,
       .           -.05566890211, .12331937 6,
       .           .18629021, .23319376,
       .           .26280454,
       .           .27292509/
        CGAUS/ ... /
        NPOINT /11/,
        IOFILE /6/,
        NTERM /4/
C
C*** Define the function FUNC and the derivative of the function DFUNC.
C*** Also define its minimum and maximum values - FMIN and FMAX.
C
```

```
      DATA FMIN, FMAX /-1., 1./
      FUNC(Z) = COS(Z)
      DFUNC(Z) = - SIN(Z)
      OPEN(IOFILE, FILE='OUTPUT.DAT', STATUS='NEW', ERR=1,
     . ACCESS='SEQUENTIAL')
      GOTO 2
1     OPEN(IOFILE, FILE='OUTPUT.DAT', STATUS='OLD',
     . ACCESS='SEQUENTIAL')
2     CONTINUE
C
C*** Initilize variables.
C
      DO 5 I = 1,NTERM
      DO 4 J = 1,NTERM
      EQ(I,J)=0.0
4     CONTINUE
      R(I)=0.0
5     CONTINUE
C
C*** Read the range for least square fitting.
C
      WRITE(*,*) 'Between what two values of X must '
     .'the function be smoothed?'
      READ(*,*) SMALL, BIG
      WIDTH = BIG-SMALL
      WRITE(*,'(A/A/A/A/2(F8.5,A))')
     .'What is the steplength in X?',WIDTH/300.,' and ',WIDTH/30.,')'
      READ(*,*) STEP
20    NSTEP = IFIX(WIDTH/STEP+1.1)
      IF (NSTEP.GT.301 .OR. NSTEP.LE.30) GO TO 20
C
C*** Prepare for graphics.
C
```

```
%%%%%%%%%%%%%        %%

      CALL GRAFON
      FMAX = FMAX - FMIN + .3*F
      FMAX = FMAX + .3*F
      FMIN = FMIN - .3*F
      CALL AXIS(SMALL-.1*WIDTH, BIG+.1*WIDTH, FMIN, FMAX,
     . 'X-axis', 'Y-axis',
     . 'Least Square Fitting of a Function by a Polynomial', 6, 6, 50)
      LFIRST = .TRUE.
      FMAX = FMAX + .6*F
      FMIN = FMIN - .6*F
C
C*** Calculate the values of the function at the integrating points.
C
      DO 60 N = 1,NPOINT
      X = (BIG+SMALL + XGAUS(N)*WIDTH)/2.
      F = FUNCC(X)
      CALL TRACE (X, F, 0)
      CALL CROSS
C
C*** Create equations (upper triangle only) as points are read.
C
      CALL VECTOR (NTERM, VEC, X)
      DO 60 I = 1,NTERM
      V = VEC(I)*CGAUS(N)
      R(I) = R(I) + F*V
      DO 60 J = I,NTERM
      EQ(I,J) = EQ(I,J) + V*VEC(J)
   60 CONTINUE
C
C*** The equations are solved by Gaussian reduction. R contains the
C*** coefficients of the polynomial fitting.
C
      CALL GAUSS (EQ, R, NTERM, LCRASH)
      IF (LCRASH) GOTO 1000
```

```
C*** Calculate in DR the derivative of the polynomial fitting.
C
      DO 100 I = 1,NTERM-1
         DR(I) = R(I+1)*I
100   CONTINUE
      DR(NTERM) = 0.
C
C*** Print the answers for comparison.
C
      SMALL = SMALL - 0.1*WIDTH
      BIG = BIG + 0.1*WIDTH
      X = SMALL
      WRITE(IO FILE,'(A/2(A,F8.4))')
     *'LEAST CURVE FITTING OF FUNCTION BY POLYNOMIAL',' BIG
     * Between X = ',SMALL,' and X = ',BIG
      WRITE (IOFILE,'(//7X,A,2X,A,3X,A,1X,A,3X,A,2X,A,/)')
     *'X','Y-approx','Y-real','Y-error','Y.-approx','Y.-real',
     *'Y.-error'
C
      DIFF = FMAX - FMIN
      FMIN = FMIN - .5*DIFF
      FMAX = FMAX + .5*DIFF
200   CONTINUE
      CALL VECTOR (NTERM, VEC, X)
      VALUE = 0.
      DVALUE = 0.
      DO 220 N = 1,NTERM
         VALUE = VALUE+R(N)*VEC(N)
         DVALUE = DVALUE+DR(N)*VEC(N)
```

```
220   CONTINUE
      FUN = FUNC(X)
      DFUN = DFUNC(X)
      ERROR = ABS(VALUE-FUN)
      DERROR = ABS(DVALUE-DFUN)
      WRITE (IOFILE,'(1X,F7.3,2(2(1X,F9.4),1X,E9.4))')
     .     X, VALUE, FUN, ERROR, DVALUE, DFUN, DERROR
C
      IF (.NOT. LFIRST) THEN
        IF (SVALUE.LT.FMAX .AND. SVALUE.GT.FMIN .AND.
     .      VALUE.LT.FMAX .AND. VALUE.GT.FMIN) THEN
          CALL TRACE (SX, SVALUE, 0)
          CALL TRACE (SX, VALUE, 1)
          CALL TRACE (X, VALUE, 0)
          CALL TRACE (X, FUN, 2)
        ENDIF
      ENDIF
      SX = X
      SVALUE = VALUE
      SFUN = FUN
      LFIRST = .FALSE.
C
      X = X + STEP
      IF (X.LE.BIG) GOTO 200
C
1000  CLOSE (IOFILE, STATUS='KEEP')
      CALL GRAFOF
      STOP
      END
```

```
C***  Subroutine to solve linear equations by Gaussian reduction:
C***  R[A].X = R;
C***  This subroutine is called by programs LEASQ and CURFIT.
C
      SUBROUTINE GAUSS (A, R, NVAR, LCRASH)
C
      IMPLICIT DOUBLEPRECISION (A-H, O-Z)
      LOGICAL LCRASH
      DIMENSION A(5, 5), R(5)
C
      LCRASH = .FALSE.
      PIVOT = A(1,1)
      DO 120 I = 1, NVAR-1
         DO 100 II = I+1, NVAR
            FACT = A(II,II)/PIVOT
            R(II) = R(II) - FACT*R(I)
            DO 100 J = I, NVAR
               A(II,J) = A(II,J) - FACT*A(I,J)
100      CONTINUE
C***  Check that the next pivot is not zero.
C
            I1 = I + 1
            PIVOT = A(I1,I1)
            LCRASH = PIVOT.LE. 1.0E-12
            IF(LCRASH) THEN
               WRITE(*,'(A/A/A/A,I2,A,E12.6)')
     .        'The equations cannot be solved:',
     .        'Either the points are not distinct, or the terms',
     .        'you chose are not functionally independent.',
     .        'The ',I1,'th pivot is only ',PIVOT
               RETURN
....
```

```fortran
120   CONTINUE
      ENDIF
C*** Back-substitution.
C
      DO 160 II = 1,NVAR
        I = NVAR + 1 - II
        GASH = R(I)
        IF (I.NE.NVAR) THEN
          DO 140 J = I+1,NVAR
            GASH = GASH - A(I,J)*R(J)
140       CONTINUE
        ENDIF
        R(I) = GASH/A(I,I)
160   CONTINUE
      RETURN
      END
C
C*** Subroutine to calculate the vector [1, x, x**2, x**3,...]
C
      SUBROUTINE VECTOR (NTERM, VEC, X)
C
      DIMENSION VEC(5)
C
      VEC(1) = 1
      DO 10 I = 1,NTERM-1
        VEC(I+1) = VEC(I)*X
10    CONTINUE
      RETURN
      END
```

```fortran
C***  Program for heat transient of a metal bar initially at zero
C***  degrees.
C
      PROGRAM THERM
C
      DIMENSION TEMP(21), EXTEMP(21)
      LOGICAL LCHECK
      CHARACTER TITLE*48
      DATA IOFILE /6/
      OPEN(IOFILE, FILE='OUTPUT.DAT', ACCESS='SEQUENTIAL', ERR=1,
     .     STATUS='NEW')
      GOTO2
1     OPEN(IOFILE, FILE='OUTPUT.DAT', ACCESS='SEQUENTIAL',
     .     STATUS='OLD')
2     CONTINUE
C
C***  Input heat data.
C
3     WRITE(*,3)
      FORMAT(6X,'(/)',19X,'A',30X,'B'/10X,30('-')/10X,'|',28X,'|'/
     . 10X,'|',28X,'|'/10X,'|',28X,'|'/10X,30('-')/
     . 3(10X,'|',28X,'|'/),10X,'X = A',23X,'X = B'/)
4     FORMAT(10X,(A/A),5,F7.3,19X,'=',F7.3//)
10    WRITE(*,*) 'At time t = 0, we expose end A to temperature Ta, end B to Tb',
     . 'What are A and B?(A less than B)'
      READ(*,*) A,B
      IF(A.GE.B) GOTO 10
      WRITE(*,*) 'What are Ta and Tb ?'
      READ(*,*) TA,TB
      WRITE(*,*) 'What is the thermal conductivity ?'
      READ(*,*) THMCON
      WRITE(*,*) 'What is the specific heat, per unit volume ?'
      READ(*,*) SPECHT
```

```
      WRITE (IOFILE,3) A,B
      WRITE (IOFILE,4)(4(A,F7.3/))')
   3  Temperature at end A = ',TA,
   4  Temperature at end B = ',TB,
C.....Thermal conductivity = ',THMCON,
C.....Specific heat, per unit volume = ',SPECHT

C***  Get timesteps.
12    WRITE(*,*)'How many intervals in X do you want ? (maximum 20)'
      READ(*,*) NXZ
      IF(NXZ.GT.20 .OR. NXZ.LT.1) GO TO 12

C***  Calculate the multiplicative factor for the table. Give warning
C***  about numeric instability.
      H = (B-A) / FLOAT(NXZ)
      FACT = THMCON /(H*H*SPECHT)
      WRITE (*,*)(A,F8.6,A)*)'What timestep do you want ? (Less than ',
     * 0.5/FACT,' to assure stability)'
      READ (*,*) EPS
      FACT = FACT * EPS

C
22    WRITE (*,*) 'How many timesteps ?'
      READ (*,*) NTZ
      WRITE'How many snapshot data checks do you want ? (At least one)'
C.....(Give small number as each check involves a plot)'
      READ (*,*) NDZ
      IF(NDZ.LT.1) GOTO 22
      WRITE (IOFILE,'(3(A,I3/))')
C.....Intervals in X = ',NXZ,
C.....Timestep = ',NTZ,
C.....Number of data checks = ',NDZ
```

```
C*** Initialize TEMP and EXTEMP.
C
      TEMP(1) = TA
      DO 30 I = 2,NXZ
      TEMP(I) = 0.
30    CONTINUE
      TEMP(NXZ+1) = TB
      EXTEMP(1) = TA
      EXTEMP(NXZ+1) = TB
C
C*** Initialize variables for the time marching loop.
C
      GRAFMX = AMAX1(TA,TB, 0.)
      GRAFMN = AMIN1(TA,TB, 0.)
      F = GRAFMX - GRAFMN
      GMAX = GRAFMX + .5*F
      GMIN = GRAFMN - .5*F
      CALL GRAFON
C
      IF(NDZ.GT.NTZ) NDZ = NTZ
      NSTEP = NTZ/NDZ
      XSTEP = (B-A) / FLOAT(NXZ)
      WRITE (IOFILE,*)'   TIME   TEMPERATURE ===>'
C
C*** Time marching loop.-------------------------------
C
      DO 40 NDUMMY = 0,NDZ-1
        DO 37 NT = 1,NSTEP
C
C*** Euler Integration (very approximate).
C
```

```
      TIME = EPS * FLOAT(NT + NDUMMY*NSTEP -1)
      WRITE (IOFILE, '(1X,F5.3,3X,11F10.6/9X,11F10.6)')
     *   TIME, (TEMP(I), I=1,NXZ+1)
      STORE = TEMP(1)
      LCHECK = NT.EQ.NSTEP
      IF (LCHECK) THEN
      IF WRITE(TITLE, '(A,F8.4)')
     *   'Heat Transient of a Metal Bar at Time = ', TIME
      CALL AXIS(A, B, GRAFMN, GRAFMX,
     *   'Distance along bar', 'Temperature', TITLE, 18, 11, 48)
      ENDIF
      SX = A
      DO 35 NX = 2,NXZ
      X = A + FLOAT(NX-1)*XSTEP
      PRE = STORE + TEMP(NX)
      STORE = TEMP(NX)
      TEMP(NX) = STORE + FACT*(PRE-2.*STORE+TEMP(NX+1))
      IF (ABS(TEMP(NX)).GT.1.E10) THEN
      WRITE (*,*) '!!!CRASH!!! Timestep longer than advised !'
      GOTO 100
      ENDIF
      IF (LCHECK) THEN
      EXTEMP(NX) = CHECK(A, B, TA, TB, THMCON, SPECHT, X, TIME)
      IF (TEMP(NX-1).GT.GMIN .AND. TEMP(NX-1).LT.GMAX
     *   .AND.TEMP(NX).GT.GMIN .AND. TEMP(NX).LT.GMAX) THEN
      CALL TRACE (SX, TEMP(NX-1), 0)
      CALL TRACE (X, TEMP(NX), 1)
      ENDIF
      IF (EXTEMP(NX-1).GT.GMIN .AND. EXTEMP(NX-1).LT.GMAX
     *   .AND.EXTEMP(NX).GT.GMIN .AND. EXTEMP(NX).LT.GMAX) THEN
      CALL TRACE (SX, EXTEMP(NX-1), 0)
      CALL TRACE (X, EXTEMP(NX), 2)
      ENDIF
      SX = X
      ENDIF
```

```
35    CONTINUE
      IF (LCHECK) THEN
         WRITE(IOFILE,'(A,11F10.6/9X,11F10.6)')
     *   CHECK,(EXTEMP(I),I=1,NXZ+1)
         IF (EXTEMP(NXZ).GT.GMIN.AND. EXTEMP(NXZ).LT.GMAX) THEN
            CALL TRACE (SX, EXTEMP(NXZ), 0)
            CALL TRACE (B, TB, 2)
         ENDIF
         IF (TEMP(NXZ).GT.GMIN.AND. TEMP(NXZ).LT.GMAX) THEN
            CALL TRACE (SX, TEMP(NXZ), 0)
            CALL TRACE (B, TB, 1)
         ENDIF
      ENDIF
37    CONTINUE
40    CONTINUE
100   CALL GRAFOF (IOFILE, STATUS='KEEP')
      CLOSE (IOFILE)
      STOP
      END
C
C***  Transient temperatures in a uniform bar, by Fourier series.
C***  This function has been written to provide a check of the answers to
C***  the previous program: THERM. It uses theory that has not been
C***  explained in the text.
C
      FUNCTION CHECK
     *(A, B, TA, TB, THMCON, SPECHT, XGIV, TGIV)
C
```

```fortran
      DATA PI/3.1415926536/
      FA=((TA+TB)*2./PI
      FB=((TA-TB)*2./PI
      TEMP=TA+(XGIV-A) * (TB-TA) / (B-A)
      X==PI*(XGIV-A)/(B-A)
      T==(PI/(B-A))**2. * TGIV * THMCON / SPECHT
      DO 10 I=1,100,2
      FII = FLOAT(I) * FI
      IF (TII.GT.85.) GOTO 100
      TERMA = (SIN(X*FII)/EXP(TII)) * (FA/FI)
      TEMP = TEMP + TERMA
      FJJ = FLOAT(I+1) * FJ
      TJJ = FLOAT(I+1) * FJ
      IF (TJJ.GT.85.) GOTO 100
      TERMB = (SIN(X*FJJ)/EXP(TJJ)) * (FB/FJ)
      TEMP = TEMP - TERMB
      ERR = EXP(-TII)/FI + EXP(-TJJ)/FJ
      IF (ERR.LT.1.0E-6*ABS(TEMP)) GOTO 100
10    CONTINUE
      ERROR=ABS(TA) / (EXP(1002001.*T)*501.) +
     .      ABS(TB) / (EXP(1004004*T)*1002.)
     .IF(ERROR .LT.1.E-6) GOTO 100
      WRITE(*,'(A,F7.4,A,F8.4)') ' Non convergence at time = ',TGIV,
     .      ' Position=',XGIV
     .
C
100   CHECK = TEMP
      RETURN
      END
```

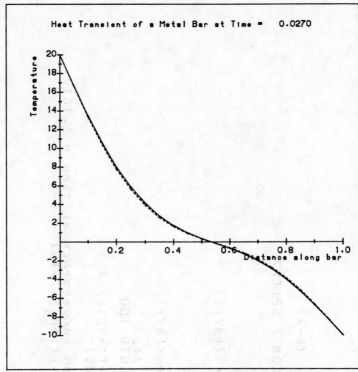

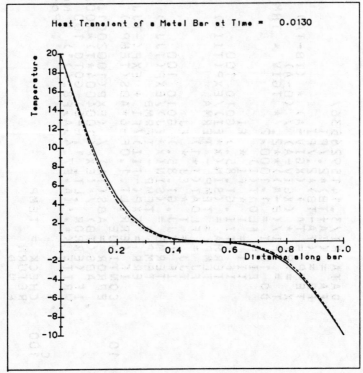

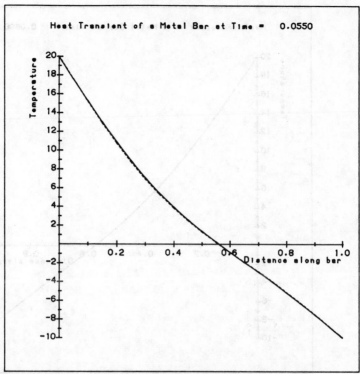

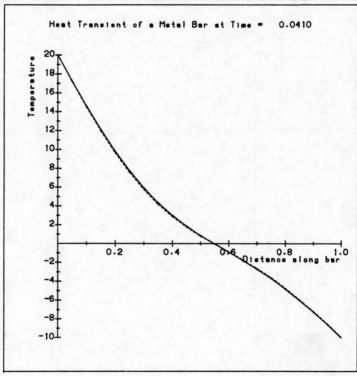

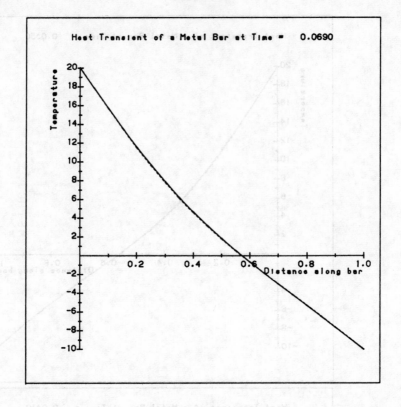

```
C*** Teaching program, for Heun's method
C     DIMENSION VAB(10), VABNEW(10), GRAD(10), PHONEY(10)
C
C***  NVAB = the number of variables excluding T.
C***  (NVAB must be less than 11)
C
      DATA NVAB /2/
C
      DATA IOFILE /6/
      OPEN(IOFILE,FILE='OUTPUT.DAT', STATUS='NEW', ERR=1,
     .     ACCESS='SEQUENTIAL')
      GOTO2
1     OPEN(IOFILE,FILE='OUTPUT.DAT', STATUS='OLD',
     .     ACCESS='SEQUENTIAL')
2     CONTINUE
C
C***  Set the initial values, x = vab(1) = ..., y = vab(2) = ..., ...
C
      VAB(1) = 0.0
      VAB(2) = 1.0
C
C***  Input data.
C
      WRITE (*,*) 'What timestep do you want ?'
      READ (*,*) EPS
      WRITE (*,*) 'How many timesteps ?'
      READ (*,*) NTZ
      WRITE(IOFILE,'(A//A,F8.5//A/)')
     . 'HEUN''S METHOD, A SECOND-ORDER RUNGE-KUTTA METHOD.',
     . ' Timestep =', EPS, ' TIME VARIABLES ===>'
C
C***  The two-step Heun integration.
C
      DO 140 NT = 1,NTZ
      TIME = EPS*FLOAT(NT-1)
```

```
      WRITE (IOFILE,'(3X,F5.3,3X,11F10.6)')
     &      TIME,(VAB(NX),NX=1,NVAB)
      DO 100 N = 1,NVAB
         PHONEY(N) = VAB(N)
100   CONTINUE
      CALL DERIV (TIME, PHONEY, GRAD)
      DO 120 N = 1,NVAB
         VABNEW(N) = VAB(N) + EPS*0.25*GRAD(N)
         PHONEY(N) = VAB(N) + EPS*0.6666667*GRAD(N)
120   CONTINUE
      TIME = TIME + 0.6666667*EPS
      CALL DERIV (TIME, PHONEY, GRAD)
      DO 140 N = 1,NVAB
         VAB(N) = VABNEW(N) + EPS*0.75*GRAD(N)
         IF (ABS(VAB(N)).GT.1.E12) THEN
            WRITE (*,*) '!!CRASH!! Numerical instability.'
            GOTO 150
         ENDIF
140   CONTINUE
150   CLOSE (IOFILE, STATUS='KEEP')
      STOP
      END
C
C *** Subroutine to define the formulae for the derivatives.
C
      SUBROUTINE DERIV(TIME, PHONEY, GRAD)
      DIMENSION PHONEY(10), GRAD(10)
C
C     Insert the derivatives here, e.g. x' = y, y' = -x
C
      GRAD(1) = PHONEY(2)
      GRAD(2) = -PHONEY(1)
      RETURN
      END
```

```
C*** Teaching program, for the fourth order Runge-Kutta method.
C       DIMENSION VAB(10), VABNEW(10), GRAD(10), PHONEY(10)
C
C*** NVAB = number of variables excluding T (NVAB must be less than 11)
C
C       DATA NVAB /1/
C
        DATA IOFILE /6/
        OPEN (IOFILE, FILE='OUTPUT.DAT', STATUS='NEW', ERR=1,
       .     ACCESS='SEQUENTIAL')
        GOTO2
1       OPEN (IOFILE, FILE='OUTPUT.DAT', STATUS='OLD',
       .     ACCESS='SEQUENTIAL')
2       CONTINUE
C
C*** Input data.
C
        WRITE (*,*)  'What timestep do you want ?'
        READ (*,*) EPS
        WRITE (*,*) 'How many timesteps ?'
        READ (*,*) NTZ
20      WRITE (IOFILE, '(A//A,F8.5/A)')
       . 'THE RUNGE-KUTTA FOURTH ORDER METHOD.',
       . ' Timestep =',EPS,' TIME VARIABLES ===>'
C
C*** Define the starting conditions.
C*** For example, x = vab(1) = 1.
C
        VAB(1) = 1.0
C
C*** Time marching loop.
C
        DO 200 NT = 1,NTZ
        TIME = EPS*FLOAT(NT-1)
        DO 15 N = 1,NVAB
        PHONEY(N) = VAB(N)
15      CONTINUE
        CALL DERIV (TIME, PHONEY, GRAD)
        DO 120 N = 1,NVAB
        VABNEW(N) = VAB(N) + EPS*GRAD(N)/6.0
```

```fortran
120     PHONEY(N) = VAB(N) + EPS*0.5*GRAD(N)
        CONTINUE
        TIME=TIME +0.5*EPS
        CALL DERIV (TIME, PHONEY, GRAD)
        DO 140 N=1,NVAB
        VABNEW(N) = VAB(N) + EPS*GRAD(N)/3.0
        PHONEY(N) = VAB(N) + EPS*0.5*GRAD(N)
140     CONTINUE
        CALL DERIV (TIME, PHONEY, GRAD)
        DO 160 N=1,NVAB
        VABNEW(N) = VABNEW(N) + EPS*GRAD(N)/3.0
        PHONEY(N) = VAB(N) + EPS*GRAD(N)
160     CONTINUE
        TIME = TIME +0.5*EPS
        CALL DERIV (TIME, PHONEY, GRAD)
        DO 180 N=1,NVAB
        VAB(N) = VABNEW(N) + EPS*GRAD(N)/6.0
        IF (ABS(VAB(N)).GT.1.E12) THEN
        WRITE (*,*) '!!CRASH!! Numerical instability.'
        GOTO 300
        ENDIF
180     CONTINUE
        WRITE(IOFILE,'(4X,F5.3,3X,11F10.6)')
        TIME,(VAB(NX), NX = 1,NVAB)
200     CONTINUE
        CLOSE (IOFILE, STATUS='KEEP')
300     STOP
        END
C
C*** Subtoutine to define the formulae for the derivatives.
C
        SUBROUTINE DERIV (TIME, PHONEY, GRAD)
        DIMENSION PHONEY(10),GRAD(10)
C*** Insert the derivatives here, e.g. x' = x + t * (x**2).
C
        GRAD(1) = PHONEY(1) + TIME*PHONEY(1)**2
        RETURN
        END
```

```fortran
C***  To invert the positive definite, symmetric matrix A by part
C***  inversion.
C
      PROGRAM POSIN
C
      DIMENSION A(6,6), VEC(6), B(6,6), C(6,6)
      CHARACTER FORMT*30
      DATA IOFILE /6/
      OPEN(IOFILE,FILE='OUTPUT.DAT', STATUS='NEW', ERR=1,
     .     ACCESS='SEQUENTIAL')
      GOTO 2
1     OPEN(IOFILE,FILE='OUTPUT.DAT', STATUS='OLD',
     .     ACCESS='SEQUENTIAL')
2     CONTINUE
C
C***  Create run-time format.
C
      WRITE(*,*)
     .'For your N X N matrix - what is N ? (between 2 and 6)'
      READ(*,*) N
      IF(N.LT.2.OR.N.GT.6) GOTO 3
      WRITE(FORMT,'(''('',I1,A)') ('','',N,''(F9.5,1X),'')'')'
C
C***  Read data. Matrix A is copied to B.
C
      WRITE(IOFILE, '(A,A//A)')
     .'INVERTING A POSITIVE DEFINITE, SYMETRIC MATRIX BY PART',
     .'...INVERSION.', 'This is the matrix that you submitted:'
      DO 10 I=1,N
```

```fortran
      DO 5 J = I,N
      WRITE (*,'(2(A,I1))') ' Enter the number in row ',I,' column ',J
      READ (*,*) AA
      A(I,J) = AA
      A(J,I) = AA
      B(I,J) = AA
      B(J,I) = AA
5     CONTINUE
      WRITE (IOFILE, FORMT) (A(I,J),J = 1,N)
10    CONTINUE
C
C*** Invert A by symmetric part-inversion
C
      DET = 1.
      DO 200 K = 1,N
      PIVOT = A(K,K)
      IF (PIVOT.LE.0.) THEN
        WRITE (*,'(A,I1,A,F9.4,A)') ' Pivot ',K,' = ',PIVOT,'. Matrix is not positive definite.'
        CLOSE (IOFILE, STATUS='KEEP.')
        STOP
      ENDIF
      DET = DET*PIVOT
      DO 140 I = 1,N
      VEC(I) = A(K,I)
140   CONTINUE
      DO 180 I = 1,N
      FACT = VEC(I)/PIVOT
      IF (I.NE.K) THEN
        DO 160 J = 1,N
        A(I,J) = A(I,J) - FACT*VEC(J)
```

```
160     CONTINUE
        ENDIF
        A(I,K) = FACT
        A(K,K) = FACT
180     CONTINUE
200     CONTINUE
        A(K,K) = -1.0/PIVOT
C
        WRITE (IOFILE,'(//A)') ' The minus inverted matrix is:'
        DO 210 I = 1,N
        WRITE (IOFILE,FORMT) (A(I,J),J=1,N)
210     CONTINUE
C
C**  As a test, calculate A.B to give minus the identity matrix?
C
        WRITE(IOFILE,'(//A)') ' The test product is:'
        DO 218 I = 1,N
        DO 216 J = 1,N
        GASH=0.
        DO 214 K = 1,N
        GASH = GASH + B(I,K)*A(K,J)
214     CONTINUE
        C(I,J) = GASH
216     CONTINUE
        WRITE (IOFILE, FORMT) (C(I,J),J=1,N)
218     CONTINUE
        WRITE (IOFILE,*)
     .  'If this is not the minus identity matrix, beware...roundoff.'
        CLOSE (IOFILE, STATUS='KEEP')
        STOP
        END
```

```
C*** Teaching program for eigenvalues.
C
      PROGRAM EIG
C
      DIMENSION A(5,5),PRE(5),POST(5)
      CHARACTER FORMT1*23,FORMT2*47
      DATA IOFILE /6/
      OPEN (IOFILE,FILE='OUTPUT.DAT', STATUS='NEW', ERR = 1,
     .      ACCESS='SEQUENTIAL')
      GOTO 2
1     OPEN (IOFILE,FILE='OUTPUT.DAT',STATUS='OLD',
     .      ACCESS='SEQUENTIAL')
2     CONTINUE
C
C***  An Eigenvalue may be:
C***  A principal stress.
C***  A principal moment of inertia in rigid body mechanics.
C***  Ditto, in 2D, for bending of beams.
C***  A time constant in a chemical reaction.
C***  A frequency or time constant in a control equation.
C***  A principal axis of an ellipse or an ellipsoid.
C***  A (frequency) squared, a buckling load, etc.
C
C***  Create run-time formats. Unless you love computing skip this
C***  paragraph.
C
10    WRITE (*,*)
     .'For your N x N matrix A - what is N ? (between 2 and 5)'
      READ (*,*) N
      IF (N.LT.2 .OR. N.GT.5) GOTO 10
C
      WRITE (FORMT1,'(A,I1,A)') '(',N,'(F9.5,1X),'')')'
      WRITE (FORMT2,'(A,2(I1,A))') '(','',N,''',''',N, ===> '(F9.5,1X),'')')'
C
```

```
C*** Input data. PRE is initialized to [1, 1, 1, ... 1]
C*** The vector PRE is initialized to [1, 1, 1, ... 1]
C
      WRITE (IOFILE,'(A//A)')
     . ' FINDING THE EIGENVALUES OF A MATRIX BY THE POWER METHOD',
     . ' This is the matrix that you submitted:'
      DO 60 I = 1,N
        DO 50 J = 1,N
          WRITE (*,'(2(A,I1))')
     .    ' Please enter the number in row ',I,' column ',J
          READ (*,*) A(I,J)
          A(J,I) = A(I,J)
50      CONTINUE
        PRE(I) = 1
      WRITE (IOFILE,FORMT1) (A(I,J), J=1,N)
60    CONTINUE
C
C*** Start of powering loop.--------------------------------------
C
      DO 280 MODES = 1,N
      WRITE(IOFILE,'(//A/)')
     . ' Vector before iteration ===> Vector after iteration'
C
C*** Power the vector, i.e. multiply by the given matrix.
C
      DO 200 ITER = 1,100
        POPR = 00.0
        POPO = 00.0
        DO 140 I = 1,N
          GASH = 0.0
          DO 120 J = 1,N
            GASH = GASH + A(I,J)*PRE(J)
```

```fortran
120   CONTINUE
      POST(I) = GASH + GASH*PRE(I)
      POPR = POPR + GASH*PRE(I)
      POPO = POPO + GASH*GASH
140   CONTINUE
C***  Write the vectors PRE and POST to the output file.
C***  We use the run-time format FORMT2 to write these vectors.
C
      WRITE (IOFILE, FORMT2) (PRE(J),J=1,N),(POST(J),J=1,N)
C
C***  Normalise the vector, and test for final convergence.
C
      IF (ABS(POPR).LT.1.E-12) THEN
        WRITE (*,*) 'Infinite eigenvalue. Try another matrix.'
        GOTO 1000
      ENDIF
      IF (ABS(POPO).LT.1.E-12) THEN
        WRITE (*,*) 'Zero eigenvalue. Try another matrix.'
        GOTO 1000
      ENDIF
      EIGEN = POPO/POPR
      FACT = 1.0/SQRT(POPO)
      IF (POPR.LT.0.0) FACT = -FACT
      ERRSQ = 0.0
      DO 160 I = 1,N
        GASH = PRE(I)
        PRE(I) = FACT*POST(I)
        ERRSQ = ERRSQ + (GASH-PRE(I))**2
160   CONTINUE
C
      IF (ERRSQ.LT.1.E-6) GO TO 220
```

```
200   CONTINUE
C
      WRITE (IOFILE,*)
     .'**Pseudo convergence stopped after 100 iterations**'
     .GOTO 1000
C
C***  "Zoo" the matrix, according to hotelling.
C
220   WRITE (IOFILE,'(A//A)')
     .'**Convergence**.','  The final normalized vector is:'
      WRITE (IOFILE,FORMT1) (PRE(I), I=1,N)
      WRITE (IOFILE,'(//A,I1,6,A/A//A)')
     .'... Zoo the matrix by Hotelling''s deflation, so that the ',
     .'MODES.-th eigenvalue, ','EIGEN,', ' becomes zero and hence offers no',
     .' just found, namely ','competition in the powering process.',
     .'     The zooed matrix is:'
C
      DO 260 I = 1,N
      FACT = PRE(I)*EIGEN
      DO 240 J = 1,N
      A(I,J) = A(I,J) - FACT*PRE(J)
240   CONTINUE
      WRITE(IOFILE, FORMT1) (A(I,J), J = 1,N)
260   CONTINUE
280   CONTINUE
C
      WRITE (IOFILE,*)
     .'If the final zooed matrix is not null, beware..roundoff.'
1000  .CLOSE (IOFILE, STATUS='KEEP')
      STOP
      END
```

```fortran
C***  Teaching program, to introduce Gauss-Seidel relaxation methods.
C***
C***  CURVE is the second derivative of W, and is used to approximate the
C***  second DIFFERENCE at successive internal points, K = 1,2,...N.
C***      CURVE = W(K-1) - 2*W(K) + W(K+1).
C***
C***  In the first step, this equation is solved for W(K):
C***      W(K)(new) = .5*(W(K-1) + W(K+1) - CURVE(K)
C***
C***  In the more sophisticated technique, the CHANGE in W(K) is amplified
C***  by the over-relaxation factor OVER:
C***      W(K)(very new) = (OVER-1)*W(K)(old) + OVER*W(K)(new).
C

      PROGRAM RELAX

      DIMENSION W(20), CURV(20)
      DATA IOFILE /6/
C
C***  N is the number of values to be calculated, including end-points
C
      DATA N /9/
C
C***  Define the function to be sought.
C***  The second derivative of the function is also defined at this point.
C
      FUNC(X) = SIN(.4*X)
      DDFUNC(X) = -.16*SIN(.4*X)
      OPEN(IOFILE,FILE='OUTPUT.DAT', STATUS='NEW', ERR=1,
     .     ACCESS='SEQUENTIAL')
      GOTO 2
1     OPEN(IOFILE,FILE='OUTPUT.DAT', STATUS='OLD',
     .     ACCESS='SEQUENTIAL')
```

```
2     CONTINUE
      WRITE (*,*) 'Give over-relaxation factor (Between 1 and 2)'
      READ (*,*) OVER
      WRITE (IOFILE,'(A//A,F7.5)')
     : 'GAUSS-SEIDEL RELAXATION METHODS',
     : 'Over-relaxation factor = ',OVER
C
C*** Initialize CURVE and W.
C
      DO 20 K = 1,N
        CURV(K) = DDFUNC(FLOAT(K-1))
200   CONTINUE
      W(1) = FUNC(0.)
      W(N) = FUNC(FLOAT(N-1))
      WRITE (IOFILE,'(3(/6X,A),17I6)')
     : 'Root mean','Current vector','square of','change ',
     : (K, K = 1,N)
C
C*** The iterative process.
C
      RMS = 0.
      DO 240 ITER = 1,100
        WRITE (IOFILE,'(I4,F10.6,3X,17F6.3)')
     :  ITER, RMS, (W(K), K = 1,N)
        IF (RMS.LT.1.E-4 .AND. ITER.NE.1) GOTO 260
        RMS = 0.
        DO 220 K = 2,N-1
          GASH = W(K)
          W(K) = GASH + OVER*(0.5*(W(K-1)+W(K+1)-CURV(K)) - GASH)
          RMS = RMS + (W(K)-GASH)**2
```

```
220    CONTINUE
       RMS = SQRT(RMS/FLOAT(N-2))
       IF (RMS.GT.1.E12) THEN
         WRITE (*,*) '!!CRASH!! Numerical Instability.'
         GOTO 300
       ENDIF
240    CONTINUE
       DO 280 K = 1,N
         W(K) = FUNC(FLOAT(K-1))
280    CONTINUE
       WRITE (IOFILE, '(/A,17F6.3)')
     . ' Exact values =', (W(K), K = 1,N)
300    CLOSE (IOFILE, STATUS='KEEP')
       STOP
       END
```

```fortran
C*** Teaching program for spline interpolation.
C       PROGRAM SPLINE
C
        DIMENSION XGIV(21), FGIV(20), DIAG(20), ABOVE(20), RHS(21)
        CHARACTER*7 POINT
        LOGICAL LFIRST
        DATA IOFILE /6/
C
C*** Define the function. Alternatively, write a function subroutine.
C*** Also define minimum and maximum values for the function
C
        DATA FMIN, FMAX /-1., 1./
        FUNC(Z) = COS(Z)
        OPEN(IOFILE,FILE='OUTPUT.DAT', STATUS='NEW', ERR=1,
     .    ACCESS='SEQUENTIAL')
        GOTO2
1       OPEN(IOFILE, FILE='OUTPUT.DAT', STATUS='OLD',
     .    ACCESS='SEQUENTIAL')
2       CONTINUE
        LFIRST = .TRUE.
C
C*** Input data.
C
10      WRITE (*,*)
     .    'How many points for spline interpolation ?',
        READ (*,*) N
        IF(N.GT.20 .OR. N.LT.2)GOTO 10
C
        BIG = 0.
        WRITE (*,*) 'Enter X values in increasing order.'
        DO 80 I = 1,N
```

```fortran
 60      WRITE (*,'(A,I2),') ' Give X coordinate of point ',I
         READ (*,*) XGIV(I)
         IF (I.NE.1) THEN
            DEL = XGIV(I) - XGIV(I-1)
            IF (DEL.GT.BIG) BIG = DEL
            IF (DEL.LE.0.) THEN
               WRITE(*,*,*)
     .         ' X-values must be in increasing order. Try again.'
               GO TO 60
            ENDIF
         ENDIF
 80      FGIV(I) = FUNC(XGIV(I))
 100     CONTINUE
         START = XGIV(1) - 2.*BIG
         FIN = XGIV(N) + 2.*BIG
         WIDTH = FIN - START
         WRITE (*,'(2(A,F8.5),A)')
     .   ' Give step length for table. (Between ',WIDTH/300.,' and ',
     .     WIDTH/30.,')'
         READ(*,*) STEP
         NSTEP = IFIX(WIDTH/STEP + 1.1)
         IF (NSTEP.GT.301 .OR. NSTEP.LE.30) GO TO 100
C
C***  Create and reduce equations - as in finite elements.
C
         DO 140 I = 1,N-1
            SPAN = XGIV(I+1) - XGIV(I)
            DIAG(I) = DIAG(I) + 4.0/SPAN
            ABOVE(I) = 2.0/SPAN
            FACT = ABOVE(I)/DIAG(I)
            DIAG(I+1) = 4.0/SPAN - FACT*ABOVE(I)
            GASH = 6.0*(FGIV(I+1)-FGIV(I))/(SPAN*SPAN)
            RHS(I+1) = RHS(I) + GASH
            RHS(I+1) = GASH - FACT*RHS(I)
 140     CONTINUE
```

```fortran
C
C*** and back-substitute.
C
      DO 160 II = 1,N
         I = N+1-II
         RHS(I) = (RHS(I) - ABOVE(I)*RHS(I+1))/DIAG(I)
160   CONTINUE
C
C*** Calculate resulting curve fitting.
C
      CALL GRAFON
      WIDTH = FMAX - FMIN
      FMIN = FMAX - 0.3*WIDTH
      FMAX = FMAX + 0.3*WIDTH
      CALL AXIS(START,FIN,FMIN,FMAX,'X-axis','Y-axis',
     .     'Spline interpolation of Given Points',6,6,36)
      FMIN = FMIN + 0.5*WIDTH
      FMAX = FMAX - 0.5*WIDTH
      INT = 0
      WRITE(IOFILE,'(A//A/)')
     .   'EXPERIMENTS WITH SPLINE INTERPOLATION.',
     .   ' X          Y-approx   Y-real     Y-error'
C
      DO 240 NST = 1,NSTEP
         X = START + STEP*FLOAT(NST-1)
180      POINT = X
         IF (INT.NE.N)THEN
            IF (ABS(X-XGIV(INT+1)).LT.0.50001*STEP) THEN
               POINT = INT
               INT = INT + 1
               GO TO 180
            ENDIF
         ENDIF
C
C*** Calculate the value, for the cubic spline.
C
```

```
      IF (INT .EQ. 0) THEN
        VALUE = FGIV(1) + RHS(1)*(X-XGIV(1))
      ELSEIF (INT.EQ.N) THEN
        VALUE = FGIV(N) + RHS(N)*(X-XGIV(N))
      ELSE
        SPAN = XGIV(INT+1) - XGIV(INT)
        XX=((X-XGIV(INT))/SPAN)
        VALUE = FGIV(INT))*(1.0-XX) + FGIV(INT+1)*XX +
     +    ((RHS(INT)-RHS(INT+1))*SPAN*0.5 + (FGIV(INT)-FGIV(INT+1)
     +    + (RHS(INT)+RHS(INT+1))*SPAN*0.5)*(1.0-XX-XX))*XX*(1.0-XX)
      ENDIF
C
C*** And print the answers for comparison.
C
      FUN = FUNC(X)
      ERROR = VALUE-FUN
      WRITE (IOFILE,'(1X,A,F5.2,1X,2(F12.6,1X),E12.6)')
     +    POINT,X,VALUE,FUN, ERROR
      IF (.NOT.LFIRST) THEN
        IF (SVALUE.GT.FMIN .AND. SVALUE.LT.FMAX .AND. VALUE.GT.FMIN
     +      .AND. VALUE.LT.FMAX) THEN
          CALL TRACE (SX, SVALUE, 0)
          CALL TRACE (X, VALUE, 1)
        ENDIF
        CALL TRACE (SX, SFUN, 0)
        CALL TRACE (X, FUN, 2)
      ENDIF
      LFIRST = .FALSE.
      SX = X
      SVALUE = VALUE
      SFUN = FUN
240   CONTINUE
C
      DO 250 I = 1,N
        CALL TRACE (XGIV(I), FGIV(I), 0)
        CALL CROSS
```

```fortran
250   CONTINUE
      CLOSE (IOFILE, STATUS='KEEP')
      CALL GRAFOF
      STOP
      END
C     Subroutine AXIS draws the axes for a graph and calculates the scale.
C     Input variables: XMIN, XMAX define the minimum and maximum values
C                      on the X-axis.
C                      Idem for YMIN and YMAX on the Y-axis.
C                      XMESS and YMESS are the labels to be attached to
C                      the X- and Y-axes.
C                      TITLE is the title of the graph.
C                      IXMESS, IYMESS and ITITLE give the length in
C                      characters of these messages.
C
      SUBROUTINE AXIS (XMIN,XMAX, YMIN, YMAX, XMESS, YMESS, TITLE,
     .IXMESS, IYMESS, ITITLE)
      CHARACTER XMESS*(*), YMESS*(*), TITLE*(*), XFORMT*20, YFORMT*20,
     .NUM*20
      LOGICAL LXINT, LYINT
      COMMON /CTRACE/ XORIG, XSCALE, YORIG, YSCALE
      COMMON /CAXIS/ XCHAR, YCHAR, XVCHAR, YVCHAR, XLEN, YLEN
C********************************************************************
C     Format variables: IXNUM, IYNUM = the maximum number of characters
C                       allowed to express the graduations.
C                       XGAP, YGAP = the minimum gap between graduation
C                       numbers expressed in centimeters.
C********************************************************************
      DATA IXNUM, IYNUM, XGAP, YGAP /7, 7, .30, .20/
C
```

```
C*** Get ready for a new graph.
C
      CALL NEWGRF (TITLE, ITITLE)
C
C*** Draw a vertical axis for the graph.
C
C*** First calculate the scales for the X- and Y-axes.
C
      WIDTH = YCHAR *
      CALL SCALE (YLEN, YMIN, YMAX-YMIN, WIDTH,
     . IYNUM, Y, YGAP, YFORMT, LYFORM, YFACT, YORIG, YSCALE,
     . YRSTEP, NYSTEP, YGSTEP, LYINT, NYLINE, YSSTEP)
C
      WIDTH = XCHAR * FLOAT(IXNUM)
      CALL SCALE (XLEN, XMIN, XMAX-XMIN, WIDTH, IXNUM, 'X', XGAP,
     . XFORMT, LXFORM, XFACT, XORIG, XSCALE, XRSTEP, NXSTEP,
     . XGSTEP, LXINT, NXLINE, XSSTEP)
C
C*** Then draw the Y-axis.
C
      YBASE = 0.
      IF(XMIN.LT.0. .AND. XMAX.GT.0.) YBASE = -XRSTEP*XORIG/XGSTEP
      XBASE = 0.
      IF(YMIN.LT.0. .AND. YMAX.GT.0.) XBASE = -YRSTEP*YORIG/YGSTEP
      CALL DRAW(YBASE,0., 'ABS', 0)
      CALL DRAW (0., YLEN, 'REL', 1)
C
C*** Write the numbers on the Y-axis.
C
      DELTA = XCHAR / 2.
      DO 10 NS = 0, NYSTEP
      FNS = FLOAT(NS)
      Y = FNS * YRSTEP
      ANUM = (YORIG + FNS*YGSTEP) * YFACT
```

```
      IF (ABS(ANUM).GT..5*YGSTEP*YFACT) THEN
        CALL DRAW (YBASE+DELTA,Y,'ABS',0)
        IF (LYINT) THEN
          WRITE (NUM,YFORMT) IFIX (ANUM)
        ELSE
          WRITE (NUM,YFORMT) ANUM
        ENDIF
        CALL GETO (NUM)
        CALL RITER (YBASE,- DELTA - XCHAR*FLOAT(LYFORM), Y+YCHAR*.5,
     '          LYFORM, NUM, 'H')
      ENDIF
      DO 10 NSS = 1,NYLINE-1
        IF (NS.EQ.NYSTEP) GOTO 10
        CALL DRAW (YBASE+DELTA/2., Y + FLOAT(NSS)*YSSTEP,
     '          'ABS',0)
        CALL DRAW (-DELTA, 0., 'REL', 1)
10    CONTINUE
C*** Label the Y-axis.
C
      CALL RITER (YBASE - XCHAR * (FLOAT(LYFORM)+2.5),
     '          YLEN - FLOAT(IYMESS)*XVCHAR,IYMESS, YMESS, 'V')
C
C*** Draw a horizontal axis for the graph.-----------------
C*** Draw the X-axis.
C
      CALL DRAW (0.,XBASE,'ABS',0)
      CALL DRAW (XLEN, 0.,'REL', 1)
C
C*** Write the numbers on the X-axis.
C
      DO 20 NS = 0, NXSTEP
        FNS = FLOAT (NS)
        X = FNS*XRSTEP
```

```
      ANUM = (XORIG + FNS*XGSTEP) *XFACT
      IF (ABS(ANUM).GT..5*XGSTEP*XFACT) THEN
      CALL DRAW(X, XBASE + DELTA,'ABS', 0)
      IF(LXINT) THEN
      WRITE (NUM,XFORMT) IFIX (ANUM)
      ELSE
      WRITE (NUM,XFORMT) ANUM
      ENDIF
      CALL RITER (X - XCHAR*FLOAT(LXFORM)*.5, XBASE - 2.*DELTA,
     .   LXFORM, NUM, 'H')
      ENDIF
      CALL GETO(NUM)
      DO 20 NSS = 1, NXLINE-1
      IF (NS.EQ.NXSTEP) GOTO 20
      CALL DRAW(X + FLOAT(NSS)*XSSTEP, XBASE + DELTA/2.,
     .   'ABS', 0) -DELTA, 'REL', 1)
   20 CONTINUE
C*** Label the X-axis.
C
      CALL RITER (XLEN - FLOAT(IXMESS)*XCHAR, XBASE -1.5*YCHAR,
     .   IXMESS, XMESS, 'H')
C
      RETURN
      END
C
C     SUBROUTINE GETO (NUM)
C
C  Subroutine to ensure that a zero precedes the decimal point.
C
      CHARACTER NUM*(*)
      DO 10 I = 1,1000
      IF (NUM(I:I).NE.'.') GOTO 15
```

```
10    CONTINUE
15    CONTINUE
      IF(NUM(I:I).EQ.' ')THEN
      NUM(I-1:I-1)='0.'
      ELSEIF (NUM(I:I+1).EQ.'0.') THEN
      NUM(I-1:I+1)='0.'
      ENDIF
      RETURN
      END

      SUBROUTINE SCALER (ALEN, AMIN, UNITS, WIDTH, NUMNUM, XY, GAP,
     .FORMT, LFORMT, FACT, AXMIN, SCALE, RSTEP, NSTEP, GSTEP,
     .LINT, NLINE, SRSTEP)
C     To calculate the scale of an axis.
C*** Input variables: ALEN = the length of the axis in centimeters.
C***                  AMIN = the lowest value to be plotted.
C***                  UNITS = the range in raster units of the
C***                          axis.
C***                  WIDTH = the width in raster units of the
C***                          graduation number.
C***                  NUMNUM = maximum number of digits permitted in
C***                           the graduation number.
C***                  XY = a character variable, = either 'X' or 'Y'
C***                       depending on the axis that is being drawn.
C***                  GAP = the gap in raster units between the
C***                        graduation numbers.
C***
C*** Output variables: FORMT = a text format for writing the numbers on
C***                           the axis.
C***                   LFORMT = the maximum number of digits in the
C***                            graduation numbers.
C***                   FACT = if the graduation numbers are expressed
C***                          exponentially, FACT = 10**-exponential.
```

```fortran
C****** AXMIN = the value of the first graduation on the
C******         axis.
C****** SCALE = the scale i.e. raster units per graph
C******         unit.
C****** RSTEP = the number of raster units per graduation.
C****** GSTEP = the number of graph units per graduation.
C****** NSTEP = the number of graduations on the axis.
C****** LINT = logical variable, .TRUE. when graduation
C******        numbers are integer values.
C****** NLINE = the number of subdivisions for each
C******         graduation (2, 5 or 10)
C****** SRSTEP = the number of raster units per sub
C******          graduation.
C
      CHARACTER FORMT*20, EXP*6, XY*1
      LOGICAL LINT
C
      IORDER (X) = NINT (ALOG10 (ABS(X)) - 0.499999)
      NUMSAV = NUMNUM
      UNSAV = UNITS
10    LINT = .FALSE.
      FACT = 1.
      IEXP = 0.
      NGP = IFIX (ALEN / (WIDTH+GAP))
      SAV = -1
      UNITS = UNSAV
C****  Calculate IGORD = the power of the graduations.
C****  We also calculate GSTEP and NLINE.
C****  NGP = the number of graduations that can be fitted into the axis.
12    UN = UNITS / FLOAT (NGP)
      IGORD = IORDER (UN) + 1.
      SUB = 10 ** (IGORD-1) 1.
      NLINE = 10
```

```fortran
      GSTEP = 10. ** IGORD
      IF(SUB*2..GT.UN) THEN
        NLINE = 2
        GSTEP = SUB * 2.
        IGORD = IGORD - 1
      ELSEIF (SUB*5..GT.UN) THEN
        NLINE = 5
        GSTEP = SUB * 5.
        IGORD = IGORD - 1
      ENDIF
      IF(GSTEP.EQ.SAV) GOTO 14
C
      AXMIN = FLOAT (NINT (AMIN / GSTEP - 0.499999)) * GSTEP
      AXMAX = FLOAT (NINT ((AMIN+UNITS) / GSTEP + 0.499999)) * GSTEP
      UNITS = AXMAX - AXMIN
      SAV = GSTEP
      GOTO 12
   14 CONTINUE
C
C**** find number of digits necessary to express numbers.
C**** ISIGN = number of characters needed for sign(0 or 1).
C**** IPOINT = (0,1 or 2)-number of characters needed for decimal point.
C****
C**** IEXP = number of characters needed for exponential value.
C**** MAXDIG = number of digits in graduation number.
C**** NUMDIG = number of digits that can be fitted into graduation
C****          number.
C
      I1 = -99999
      I2 = -99999
      IF (AXMIN.NE.0.) I1 = IORDER (AXMIN)
      IF (AXMAX.NE.0.) I2 = IORDER (AXMAX)
      IORD = MAX0 (I1, I2)
      ISIGN = 0
      IF (AXMIN.LT.0.) ISIGN = 1
```

```
      IF(I1.LT.I2 .AND. AXMAX.GT.0. .AND. I2.GT.0) ISIGN = 0
      IPOINT=0
      IF(IGORD.LT.0) IPOINT = 1
      IF(IORD.LT.0) IPOINT = 2
      MAXDIG=IGORD +1
      IF(IGORD.LT.0) MAXDIG = IORD - IGORD + 1
      IF(IORD.LT.0) MAXDIG = -IGORD
      NUMDIG = NUMNUM - ISIGN - IPOINT
C
C*** See if exponential is needed.
C
      IF (MAXDIG.GT.NUMDIG .AND.
     . ((IORD.GT.0.AND.IGORD.GT.0) .OR. (IORD.LT.0.AND.IGORD.LT.0))
     . THEN
      IPOINT = 1
      IF(AXMIN.LT.0.) ISIGN = 1
      NUMDIG=NUMNUM-ISIGN-IPOINT
      MAXDIG=IGORD+1-IPOINT
      IEXP=IABS(IORDER(FLOAT(IORD))) + 2
      IF(IORD.LE.0) IEXP = IEXP + 1
      NUMDIG=NUMDIG-IEXP
      FACT=10.**(-IORD)
      ENDIF
C
C*** Test for overflow of graduation numbers.
C
      IF (MAXDIG.GT.NUMDIG+NUMSAV-NUMNUM) WRITE(*,'(4A,I2,A)')
     . '**WARNING -- The numbers on the ',XY,'-axis are ',
     . 'longer than ',NUMSAV,'.'
C
C*** Calculate LFORMT, the number of characters in the graduation
C*** numbers and FORMT, the format for the graduation numbers.
C
      LFORMT = MAXDIG + ISIGN + IPOINT + IEXP
```

```
      IF (LFORMT .LT. NUMNUM-1 .AND. XY .EQ. 'X') THEN
         WIDTH = (WIDTH/NUMNUM) * FLOAT (NUMNUM-1)
         NUMNUM = NUMNUM - 1
         GOTO 10
      ENDIF
      IF (IEXP .EQ. 0) THEN
         IF (IPOINT .EQ. 0) THEN
            WRITE (FORMT,100) LFORMT
100         FORMAT ('('I2,'.')')
            LINT = .TRUE.
         ELSE
            WRITE (FORMT,102) LFORMT, -IGORD
102         FORMAT ('(F',I2,'.',I2,')')
         ENDIF
      ELSE
         WRITE (FORMT,104) IEXP-1
104      FORMAT ('('E'I'.'I2,'E',I2,')')
         WRITE (EXP,FORMT) IORD
         WRITE (FORMT,106) MAXDIG+1, MAXDIG-1, EXP
106      FORMAT ('(F',I2,'.',I2,',',A6,')')
      ENDIF
C
C***  Calculate remaining output variables.
C
      SCALE = ALEN / UNITS
      RSTEP = GSTEP * SCALE
      SRSTEP = RSTEP / FLOAT(NLINE)
      NSTEP = NINT (UNITS/GSTEP)
C
      NUMNUM = NUMSAV
      UNITS = UNSAV
      RETURN
      END
```

```
C
C***  Subroutine TRACE moves the pen (up or down, absolute) to the point
C***  X, Y expressed in graph coordinates.
C
      SUBROUTINE TRACE (X, Y, UPDOWN)
      COMMON /CTRACE/ XORIG, XSCALE, YORIG, YSCALE
      XNEW = (X - XORIG) * XSCALE
      YNEW = (Y - YORIG) * YSCALE
      CALL DRAW (XNEW, YNEW, 'ABS', UPDOWN)
      RETURN
      END
C
C***  Subroutine CROSS draws a cross at the pen position.
C***  The user must define CROSIZ - the radius length of the cross in
C***  raster units.
C
      SUBROUTINE CROSS
      DATA CROSIZ /.2/
      CALL DRAW (CROSIZ, CROSIZ, 'REL', 0)
      CALL DRAW (-2.*CROSIZ, -2.*CROSIZ, 'REL', 1)
      CALL DRAW (0.,2.*CROSIZ, 'REL', 0)
      CALL DRAW (2.*CROSIZ, -2.*CROSIZ, 'REL', 1)
      CALL DRAW (-CROSIZ, CROSIZ, 'REL', 0)
      RETURN
      END
C
C***  Graphics routines to assure portability of programs.
C***  As an example we print the routines for the CALCOMP graphics
C***  Language.
C
C***  Subroutine GRAFON turns on and initializes the graphics device.
C***  XPAPER, YPAPER give the dimensions of the plotting paper in cms.
C
```

```
      SUBROUTINE GRAFON
      LOGICAL LFIRST
      COMMON /FIRST/ LFIRST
      COMMON /PAPER/ XPAPER, YPAPER
      COMMON /CAXIS/ XCHAR, YCHAR, XVCHAR, YVCHAR, XLEN, YLEN
C
C***  Define size of characters.
C***  CHARSZ gives the dimensions of the characters in cms.
C***  (The character is square.)
C
      DATA CHARSZ /.18/
C
C***  LFIRST signals the start of plotting.
C
      LFIRST = .TRUE.
C
C***  Calculate the variables.
C
      XCHAR = CHARSZ
      YCHAR = CHARSZ * 1.5
      XVCHAR = CHARSZ
      YVCHAR = YCHAR
C
C***  Open plotting output file, and define size of plotting paper.
C
      CALL PLOTS
      CALL PAGE_SIZE (XPAPER, YPAPER)
      RETURN
      END
C
C***  Subroutine GRAFOF turns off the graphics device.
C
      SUBROUTINE GRAFOF
      CALL PLOT (0., 0., 999)
      RETURN
      END
```

```
C
C*** Subroutine NEWGRF readies the graphics device for a new graph.
C
      SUBROUTINE NEWGRF (TITLE, ITITLE)
C
      LOGICAL LPAGE, LFIRST
      CHARACTER TITLE*(*)
C
C*** Common PAPER defines the size of the graphics paper.
C*** If XPAPER or YPAPER > 1000, the paper is scrolled in that
C*** dimension.
C
      COMMON /PAPER/ XPAPER, YPAPER, YPAPER
C
C*** Common CAXIS contains the variables to be used in the routine CAXIS.
C*** XCHAR, YCHAR give the dimensions of the characters in cms.
C*** XVCHAR, YVCHAR give the dimensions of vertical characters in cms.
C*** XLEN, YLEN give the length and breadth in cms of the graph.
C
      COMMON /CAXIS/ XCHAR, YCHAR, XVCHAR, YVCHAR, XLEN, YLEN
C
C*** Common INIT contains variables which must be conserved from one
C*** call of NEWGRF to the next.
C*** Common FIRST signals the first call to NEWGRF.
C
      COMMON /INIT/ XNEXT, YNEXT, XORIGN, YORIGN, LPAGE
      COMMON /FIRST/ LFIRST
C
C*** Define size of plot in centimeters and DELT = margin between plots.
C
      DATA XSCREN, YSCREN, DELT /13., 13., .2/
C
```

```
C*** Initialize XNEXT, YNEXT = bottom left corner of plot.
C*** LPAGE signals that a new page is needed.
C*** XORIGN, YORIGN = the current graphics origin.
C
      IF (LFIRST) THEN
        LFIRST = .FALSE.
        XNEXT = DELT
        YNEXT = DELT
        XORIGN = 0.
        YORIGN = 0.
        LPAGE = .FALSE.
      ENDIF
C
C*** Draw square around graph.
C
      IF (LPAGE) THEN
        CALL PLOTS (0., 0., 999)
        XORIGN = 0.
        YORIGN = 0.
      ENDIF
      CALL PLOT (XNEXT-XORIGN, YNEXT-YORIGN, -3)
      CCALL DRAW (0., YSCREN, 'ABS', 1)
      CCALL DRAW (XSCREN, YSCREN, 'ABS', 1)
      CALL DRAW (XSCREN, 0., 'ABS', 1)
      CALL DRAW (0., 0., 'ABS', 1)
C
C*** Write title on graph.
C
      CALL RITER (.5 * (XSCREN - FLOAT(ITITLE)*XCHAR),
     .  YSCREN - 2.*YCHAR, ITITLE, TITLE, 'H')
```

```fortran
C*** Move origin to bottom left hand corner of the graph - XLEN, YLEN.
C*** and calculate the size of the actual graph - XLEN, YLEN.
C
      XOFF = 10.*XCHAR
      YOFF = 4.*YCHAR
      XORIGN = XNEXT+XOFF
      YORIGN = YNEXT+YOFF
      CALL PLOT (XOFF, YOFF, -3)
      XLEN = XSCREN - 13.*XCHAR
      YLEN = YSCREN - 10.*YCHAR
C
C*** Calculate bottom left corner - XNEXT, YNEXT - for next graph.
C
      X = XSCREN + DELT
      Y = YSCREN + DELT
      IF (XPAPER.GT.1000.) THEN
        YNEXT = YNEXT + Y
        IF ((YNEXT+Y).GT.YPAPER) THEN
          YNEXT = XNEXT + X
          XNEXT = XNEXT + X
        ENDIF
      ELSE
        XNEXT = XNEXT + X
        IF ((XNEXT+X).GT.XPAPER) THEN
          XNEXT = DELT
          YNEXT = YNEXT + Y
          LPAGE = (YNEXT+Y).GT.YPAPER
          IF (LPAGE) THEN
            YNEXT = DELT
            YNEXT = DELT
          ENDIF
        ENDIF
      ENDIF
      RETURN
      END
```

```
C
C*** Subroutine DRAW moves the pen (up or down, absolute of relative) to
C*** the point X, Y expressed in centimeters.
C
      SUBROUTINE DRAW (X, Y, TYPE, ILINE)
      CHARACTER TYPE*3
      COMMON /SAV/ XSAV, YSAV
C
C*** DLEN = the length of the dashes in the dotted line (ILINE=2).
C
      DATA DLEN /.08/
C
C*** Calculate IPEN, = 2 if pen down, 3 if pen up.
C
      IPEN = 3
      IF (ILINE.NE.0) IPEN = 2
      IF (TYPE.EQ.'REL') THEN
        GX = XSAV+X
        GY = YSAV+Y
      ELSE
        GX = X
        GY = Y
      ENDIF
C
C*** Draw the line. If ILINE = 2 draw a dotted line.
C
      IF (ILINE.NE.2) THEN
        CALL PLOT (GX, GY, IPEN)
      ELSE
        CALL DASHPT (GX, GY, DLEN)
      ENDIF
```

```
C*** Save current cursor position.
C
      XSAV = GX
      YSAV = GY
      RETURN
      END
C
C*** Subroutine RITER writes a string STRING of length ISTRIN with
C*** upper left corner of the first character at position X, Y.
C*** The string will be written vertically if VH.EQ.'V' and
C*** horizontally if VH.EQ.'H'.
C
      SUBROUTINE RITER (X, Y, ISTRIN, STRING, VH)
      CHARACTER STRING*(*), VH
      COMMON /CAXIS/ XCHAR, YCHAR, XVCHAR, YVCHAR, XLEN, YLEN
      IF (VH.EQ.'V') THEN
         ANGLE = 90.
         GX = X + XCHAR
         GY = Y - XCHAR
      ELSE
         GX = X
         GY = Y - XCHAR
      ENDIF
      CALL SYMBOL (GX, GY, XCHAR, STRING, ANGLE, ISTRIN)
      RETURN
      END
```

Answers

2.1 (a) $\xi = 0.214114$ (b) $\xi = 0.106367$

2.2 (a) $1 + 2x - \dfrac{x^3}{6}$ (b) $0.97048 + 2.65193x - 0.15059x^2 - 0.09005x^3$

2.3 (a) 1st 7 terms gives 0.8554, 11 terms 0.8556: actual 0.8555
(b) Diverges: no sensible answer.

2.4 $2x^5$

2.5 $h = 0.1; f' = -0.90778; f'' = 0.41580$
$h = 0.05; f' = -0.90892; f'' = 0.41606$
actual; $f' = -0.09030; f'' = 0.41615$

2.6 Using Taylor of order 4 means using the first four terms in Taylor's expansion (i.e. up to and including the 3rd differential)
5.956, 4.038, 2.075, -0.110, -2.079 m/s; 0.79 s.

3.1 (a) $1 + \cfrac{1}{2 + \cfrac{1}{2 + \cfrac{1}{2 + \cfrac{1}{2 + \cfrac{1}{2 + \cfrac{1}{2} \cdots}}}}}$ (b) 198:140

3.2 $\left(\left(\left((\ldots + 1)\dfrac{x^2}{56} - 1 \right)\dfrac{x^2}{30} + 1 \right)\dfrac{x^2}{12} - 1 \right)\dfrac{x^2}{2} + 1$

(Are there any conditions on the value of x for which this sort of calculation would be of use?)

3.3 $\dfrac{1}{2} + \dfrac{1}{4}\cdot\dfrac{1}{2} + \dfrac{1}{8}\cdot\dfrac{1}{3} + \cdots + \dfrac{1}{2^n}\cdot\dfrac{1}{n}$

4.2 (i) -0.00012 (ii) 10.96

4.3 Linear 1.058238, Lagrange 1.058301, actual 1.058301

4.4 $f(x) = x^2 - 3x$

4.6 $130,427$; $229,433$

4.7 (a) Using lower three temps: $5000\mu = 0.14T^2 - 35T + 355.0$; $\mu(85) = 0.32$

    (b) Using higher three temps: $5000\mu = 0.07T^2 - 24.5T + 3200$; $\mu(85) = 0.32$

4.8 (a) 626 kPa   (b) 612 or 583 kPa

5.1 $y = 4x + 4$

5.2 57.0% coarse, 37.7% sand, 5.3% fines

5.3 $f(x) = 20.51 - 4.84x + 0.99x^2$

5.4 $a = 2.117488628$, $b = -1.458850505$, $c = 0.3274004184$,
exact at $x = 1.1021076, 1.4755564, 1.8781963$

5.5 $H(t) = 0.93333 + 0.57735 \sin \dfrac{2\pi t}{12} + 0.26667 \cos \dfrac{2\pi t}{12}$

5.6 $\begin{bmatrix} h_A \\ h_B \\ h_C \end{bmatrix} = \begin{bmatrix} 1.25 \\ 1.75 \\ 3.0 \end{bmatrix}$

6.2 (i) $3.141068, 3.133333, 3.142118$
(ii) $0.693122, 0.694444, 0.693175$

6.3 (a) exact $\dfrac{\pi}{2}$; Romberg, $\dfrac{\pi}{2}$ if leave out $N = 1$, otherwise 0.4999993π at 5th
level.
(b) exact 0.3465735903; Romberg, 0.3465735917

6.4 1.812804954

6.5 $\dfrac{1}{81} [25\phi(-\sqrt{0.6}, -\sqrt{0.6}) + 40\phi(-\sqrt{0.6}, 0) + 25\phi(-\sqrt{0.6}, \sqrt{0.6})$

$+ 40\phi(0, -\sqrt{0.6}) + 64\phi(0, 0) + 40\phi(0, \sqrt{0.6}) + 25\phi(\sqrt{0.6}, -\sqrt{0.6})$

$+ 40\phi(\sqrt{0.6}, 0) + 25\phi(\sqrt{0.6}, \sqrt{0.6})]$
Integral: 1.777777.

6.6 33.741 m (to the nearest mm!!?)

6.7   2.358373589, 2.355011236, 2.354778914, 2.355011114

7.1 ± 0.96592583, ± 0.70710678, $\pm 0.25881905 : 1.000 : |r| = 1$

$(r = x \pm i\sqrt{1 - x^2})$

7.2 $f(\theta) = 1.00000 - 0.46952\theta^2$; error 0.00982

7.4   (a) 245.7, 570.4, 1029.6, 1354.3°C
      (b) 283.3, 596.0, 1004.0, 1316.7°C

7.5   (a) 296.7 and 479.7°K
      (b) 307.5 and 454.0°K

8.1   one time step 0.01   4.095   3.540   4.095
                            3.020   0.105   3.020
                            2.090   1.060   2.090
      10 time steps 0.01   4.824   3.945   4.824
                            3.276   0.942   3.276
                            2.789   1.610   2.789
      1 time step 0.1   4.95   3.90   4.95
                            3.20   1.05   3.20
                            2.90   1.60   2.90
      20 time steps 0.01   5.438   4.433   5.438
                            3.649   1.714   3.649
                            3.392   2.202   3.392
      1 time step 0.2   5.90   4.30   5.90
                            3.40   2.10   3.40
                            3.80   2.20   3.80
      30 time steps 0.01   5.920   4.908   5.920
                            4.048   2.381   4.048
                            3.873   2.747   3.873
      1 time step 0.3   6.85   4.70   6.85
                            3.60   3.15   3.60
                            4.70   2.80   4.70
      estimate of ε for instability: 0.25

8.2   (i) $1.030909 : x = e^{(t^2 - 1)/2} = 1.03090918$ *for* $t = 1.03$
      (ii) $1.030900 : x = -1 - t + 2e^t = 1.030909$ for $t = 0.03$
      (iii) 0.610690 (one step):
            0.611099 (two steps): $x = \sqrt{t^2 - 0.75} = 0.611228$ for $t = 1.06$

8.3   (i) $1.246071 : x = e^{(t^2 - 1)/2} = 1.246077$ *for* $t = 1.2$
      (ii) $1.242800 : x = 1 - t + 2e^t = 1.24806$ for $t = 0.2$
      (iii) $0.678253 : x = \sqrt{t^2 - 0.75} = 0.678233$ for $t = 1.1$

8.4   (i) $0.241333 : x = e^t - \cos t = 0.241336$ for $t = 0.2$
      (ii) $0.205333 : x = \frac{1}{2}\sinh 2t = 0.205376$ for $t = 0.2$
      (iii) $0.886200 : x = e^t - \frac{1}{2}e^{-2t} = 0.886243$ for $t = 0.2$

8.6 'Exact' solution: $\theta = -24.6227°$ (What is the size of error to be expected from the assumption $\sin \theta = \theta$? How would this affect the period calculated and what is the effect over many cycles?)
Runge–Kutta using $\Delta t = 0.5$ $\theta = -23.4055°$
       using $\Delta t = 0.01$ $\theta = -23.4210°$

9.1 Series 0.69316334; exact 0.69314718

9.2 Newton $x = 1.366025404$
Secant $x = 1.359200833$ (at 5), $x = 1.366024156$ (at 15)
exact $x = 1.366025404$

9.3 1.10 m

9.4

x	y	z
$\pm$ 2.0632228613	± 0.6334950716	-1.1807723149
$\pm$ 4.7831262441	± 0.2186242520	-3.0854951257
$\pm$ 7.8853028901	± 0.1288911484	-4.1137876750
± 11.0146823942	± 0.0915424441	-4.7901816903
± 14.1505806436	± 0.0710231703	-5.2945047706
± 29.8500686404	± 0.0335384007	-6.7912513475

9.5 $x = 1.79010567$ $y = 3.00646674$

9.6 $y_2 = 0.885$ m, $y_3 = 11.565$ m, $Z = 18.435$ m

10.1 $w = 1, x = 2, y = 3, z = 4$

10.2 $x = 1.87480, y = 0.10509, z = 1.12438$
giving check answers of 175.00011, 193.99969, 165.00036 actual answers
$x = y = z = 1$

10.3 $\dfrac{1}{54} \begin{bmatrix} 18 & 27 & -9 \\ -4 & -15 & 17 \\ 4 & -12 & 10 \end{bmatrix}$

10.5 $\begin{bmatrix} 1 & 0 & 0 & 0 \\ -1 & 1 & 0 & 0 \\ 1 & -1 & 1 & 0 \\ 1 & 1 & -\frac{3}{4} & 1 \end{bmatrix} \begin{bmatrix} 1 & 0 & 0 & 0 \\ 0 & 4 & 0 & 0 \\ 0 & 0 & 16 & 0 \\ 0 & 0 & 0 & 25 \end{bmatrix} \begin{bmatrix} 1 & -1 & 1 & 1 \\ 0 & 1 & -1 & 1 \\ 0 & 0 & 1 & -\frac{3}{4} \\ 0 & 0 & 0 & 0 \end{bmatrix}$

10.6

9	0	0	0	0	0	0
0	27.88888889	0	0	0	0	0
0	0	3.89243028	0	0	0	0
0	0	0	24.83418628	0	0	0
0	0	0	0	10.67786341	0	0
0	0	0	0	0	16.37771396	0
0	0	0	0	0	0	10.31621162

Voltages: 17.44070330, 40.22758616, 40.72639131, 36.33898954, 29.93717712, 34.5361812, 25.50371918.

11.1 (i) $18, -6, -2$ $\begin{bmatrix} 1 \\ 2 \\ -1 \end{bmatrix}$ $\begin{bmatrix} 1 \\ -1 \\ -1 \end{bmatrix}$ $\begin{bmatrix} 1 \\ 0 \\ 1 \end{bmatrix}$

Checks: (a) the eigenvectors are orthogonal;
(b) the sum of the eigenvalues = the sum of the diagonals;
(c) the product of the eigenvalues = the determinant of the matrix.

(ii) $6, 3, -2$ $\begin{bmatrix} 2 \\ -1 \\ -1 \end{bmatrix}$ $\begin{bmatrix} 1 \\ 1 \\ 1 \end{bmatrix}$ $\begin{bmatrix} 0 \\ 1 \\ -1 \end{bmatrix}$

(iii) $5, 2, 1$ $\begin{bmatrix} -1 \\ 2 \\ 1 \end{bmatrix}$ $\begin{bmatrix} 1 \\ 1 \\ -1 \end{bmatrix}$ $\begin{bmatrix} 1 \\ 0 \\ 1 \end{bmatrix}$

(iv) $-6, 3, 2$ $\begin{bmatrix} 1 \\ 2 \\ -1 \end{bmatrix}$ $\begin{bmatrix} -1 \\ 1 \\ 1 \end{bmatrix}$ $\begin{bmatrix} 1 \\ 0 \\ 1 \end{bmatrix}$

11.2 (i) $\omega^2 = \dfrac{g}{l}\left(\dfrac{11 \pm \sqrt{103}}{3}\right)$ faster frequency for $\theta_1\theta_2$ of opposite sign.

(ii) $\omega^2 = \dfrac{k}{m}\left(\dfrac{14 \pm 8}{6}\right)$

11.3 Neither matrix is positive definite.

11.4 (a) -11.61 MPa (b) -3.03 MPa (c) -2.32 MPa

at $\begin{bmatrix} 0.1789 \\ 0.1270 \\ 1.0000 \end{bmatrix}$ at $\begin{bmatrix} -0.3130 \\ 1.0000 \\ -0.0715 \end{bmatrix}$ at $\begin{bmatrix} 1.0000 \\ 0.2973 \\ -0.2123 \end{bmatrix}$

The eigenvalues are the principal stresses and the eigenvectors are the directions of the principal stresses relative to the x, y, z axes.

12.1 (a) hyperbolic
(b) parabolic
(c) hyperbolic
(d) elliptical

12.2 (i) hyperbola
(ii) hyperbola
(iii) ellipsoid
(iv) paraboloid
(v) hyperboloid
(vi) two lines
(vii) ellipse

(viii) hyperbola

(ix)   ellipse

(x)   parabola

12.3 (i)   (a)   91.6   98.7

                 59.7

        (b)   45   35

                39

    (ii)   (a)   148.1   187.6   exact 133.3   166.7

               187.6   135.0   166.7   133.3

        (b)   40   68

                68   40

13.1 (i)

$$\begin{bmatrix} \theta_1 \\ \theta_2 \\ \theta_3 \\ \theta_4 \\ \theta_5 \end{bmatrix} = \frac{1}{336} \begin{bmatrix} 97 & -26 & 7 & -2 & 1 \\ -26 & 52 & -14 & 4 & -2 \\ 7 & -14 & 49 & -14 & 7 \\ -2 & 4 & -14 & 52 & -26 \\ 1 & -2 & 7 & -26 & 97 \end{bmatrix} \begin{bmatrix} 12 \\ 0 \\ -18 \\ 6 \\ 12 \end{bmatrix}$$

$f(2.5) = 6.274553571 \rightarrow 6.3$ (to the same number of decimals in the table)

(ii)

$$\begin{bmatrix} \theta_2 \\ \theta_5 \\ \theta_9 \\ \theta_{12} \\ \theta_{17} \end{bmatrix} = \frac{1}{1836} \begin{bmatrix} 1621 & -488 & 116 & -40 & 20 \\ & 976 & -232 & 80 & -40 \\ & & 928 & -320 & 160 \\ & & & 1060 & -530 \\ & & & & 2560 \end{bmatrix} \begin{bmatrix} \frac{2}{3} \\ \frac{7}{24} \\ -\frac{3}{8} \\ \frac{6}{5} \\ \frac{6}{5} \end{bmatrix}$$

$f(15) = 5.504357299 \rightarrow 5.5$ (to the same number of decimals in the table)

13.2 $A = -1, B = -21, C = -24, D = -12.$

Index

Mathematics and its Applications

Series Editor: G. M. BELL, Professor of Mathematics, King's College (KQC), University of London

Gardiner, C.F.	Algebraic Structures: with Applications
Gasson, P.C.	Geometry of Spatial Forms
Goodbody, A.M.	Cartesian Tensors
Goult, R.J.	Applied Linear Algebra
Graham, A.	Kronecker Products and Matrix Calculus: with Applications
Graham, A.	Matrix Theory and Applications for Engineers and Mathematicians
Griffel, D.H.	Applied Functional Analysis
Griffel, D.H.	Linear Algebra*
Hanyga, A.	Mathematical Theory of Non-linear Elasticity
Harris, D.J.	Mathematics for Business, Management and Economics
Hoskins, R.F.	Generalised Functions
Hoskins, R.F.	Standard and Non-standard Analysis*
Hunter, S.C.	Mechanics of Continuous Media, 2nd (Revised) Edition
Huntley, I. & Johnson, R.M.	Linear and Nonlinear Differential Equations
Jaswon, M.A. & Rose, M.A.	Crystal Symmetry: The Theory of Colour Crystallography
Johnson, R.M.	Theory and Applications of Linear Differential and Difference Equations
Kim, K.H. & Roush, F.W.	Applied Abstract Algebra
Kosinski, W.	Field Singularities and Wave Analysis in Continuum Mechanics
Krishnamurthy, V.	Combinatorics: Theory and Applications
Lindfield, G. & Penny, J.E.T.	Microcomputers in Numerical Analysis
Lord, E.A. & Wilson, C.B.	The Mathematical Description of Shape and Form
Marichev, O.I.	Integral Transforms of Higher Transcendental Functions
Massey, B.S.	Measures in Science and Engineering
Meek, B.L. & Fairthorne, S.	Using Computers
Mikolas, M.	Real Function and Orchogonal Series
Moore, R.	Computational Functional Analysis
Müller-Pfeiffer, E.	Spectral Theory of Ordinary Differential Operators
Murphy, J.A. & McShane, B.	Computation in Numerical Analysis*
Nonweiller, T.R.F.	Computational Mathematics: An Introduction to Numerical Approximation
Ogden, R.W.	Non-linear Elastic Deformations
Oldknow, A. & Smith, D.	Learning Mathematics with Micros
O'Neill, M.E. & Chorlton, F.	Ideal and Incompressible Fluid Dynamics
O'Neill, M.E. & Chorlton, F.	Viscous and Compressible Fluid Dynamics*
Page, S. G.	Mathematics: A Second Start
Rankin, R.A.	Modular Forms
Ratschek, H. & Rokne, J.	Computer Methods for the Range of Functions
Scorer, R.S.	Environmental Aerodynamics
Smith, D.K.	Network Optimisation Practice: A Computational Guide
Srivastava, H.M. & Karlsson, P.W.	Multiple Gaussian Hypergeometric Series
Srivastava, H.M. & Manocha, H.L.	A Treatise on Generating Functions
Shivamoggi, B.K.	Stability of Parallel Gas Flows*
Stirling, D.S.G.	Mathematical Analysis*
Sweet, M.V.	Algebra, Geometry and Trigonometry in Science, Engineering and Mathematics
Temperley, H.N.V. & Trevena, D.H.	Liquids and Their Properties
Temperley, H.N.V.	Graph Theory and Applications
Thom, R.	Mathematical Models of Morphogenesis
Toth, G.	Harmonic and Minimal Maps
Townend, M. S.	Mathematics in Sport
Twizell, E.H.	Computational Methods for Partial Differential Equations
Wheeler, R.F.	Rethinking Mathematical Concepts
Willmore, T.J.	Total Curvature in Riemannian Geometry
Willmore, T.J. & Hitchin, N.	Global Riemannian Geometry
Wojtynski, W.	Lie Groups and Lie Algebras*

Statistics and Operational Research

Editor: B. W. CONOLLY, Professor of Operational Research, Queen Mary College, University of London

Beaumont, G.P.	Introductory Applied Probability
Beaumont, G.P.	Probability and Random Variables*
Conolly, B.W.	Techniques in Operational Research: Vol. 1, Queueing Systems*
Conolly, B.W.	Techniques in Operational Research: Vol. 2, Models, Search, Randomization
Conolly, B.W.	Lecture Notes in Queueing Systems

Statistics and Operational Research

Editor: B. W. CONOLLY, Professor of Operational Research, Queen Mary College, University of London

French, S.	Sequencing and Scheduling: Mathematics of the Job Shop
French, S.	Decision Theory: An Introduction to the Mathematics of Rationality
Griffiths, P. & Hill, I.D.	Applied Statistics Algorithms
Hartley, R.	Linear and Non-linear Programming
Jolliffe, F.R.	Survey Design and Analysis
Jones, A.J.	Game Theory
Kemp, K.W.	Dice, Data and Decisions: Introductory Statistics
Oliveira-Pinto, F.	Simulation Concepts in Mathematical Modelling*
Oliveira-Pinto, F. & Conolly, B.W.	Applicable Mathematics of Non-physical Phenomena
Schendel, U.	Introduction to Numerical Methods for Parallel Computers
Stoodley, K.D.C.	Applied and Computational Statistics: A First Course
Stoodley, K.D.C., Lewis, T. & Stainton, C.L.S.	Applied Statistical Techniques
Thomas, L.C.	Games, Theory and Applications
Whitehead, J.R.	The Design and Analysis of Sequential Clinical Trials

*In preparation